Terrestrial Gamebirds
& Snipes of Africa

Terrestrial Gamebirds & Snipes of Africa

Guineafowls, Francolins, Spurfowls, Quails, Sandgrouse & Snipes

Rob Little

ACKNOWLEDGEMENTS

My appreciation begins with Prof. Tim Crowe, the Percy FitzPatrick Institute of African Ornithology and the University of Cape Town for the opportunity to develop my conservation biology career with gamebird research in the early years.

I thank Nicky and Strilli Oppenheimer of E Oppenheimer & Son sincerely for their generous support towards the publication costs of the book. Without their contribution the book would undoubtedly not have happened. E Oppenheimer & Son is the investment holding company of the Oppenheimer Family interests, founders of the global mining company Anglo American, and managing shareholders of De Beers (the world's leading diamond company), until its acquisition by Anglo American. The Oppenheimer Family has a rich history and association with conservation work in South Africa, having long been supporters of biodiversity conservation and research in natural sciences and wildlife administration. Through numerous programmes, E Oppenheimer & Son remains a dedicated custodian of South Africa's unique ecological heritage.

Special thanks also to Bryan Little and Jessica da Silva for compiling the distribution maps.

A plea to bird photographers across the world resulted in an overwhelmingly generous offer of over 950 photographs from more than 60 accomplished bird photographers for consideration. A difficult task then lay ahead to select the most appropriate photographs which would represent the birds. The list on page 6 reflects the names of the photographers and the page numbers of their photographs. Special gratitude goes to Hadoram Shirihai for his photograph of the Mount Cameroon Spurfowl, which was used for the cover.

I am also grateful to Jacana Media for taking on this publishing project and who have created a high quality product which I trust is a worthy addition to the ornithological literature and to the bookshelves of bird enthusiasts, particularly birders, wing-shooters, land owners and anyone with an interest in nature and conservation, throughout Africa and across the rest of the world.

OPPOSITE Calling male Clapperton's Spurfowl (photo by Nik Borrow)

PHOTOGRAPHER LIST

Barelli, Lorenzo 24–5, 43, 133, 152, 208
Baston, Bill 97
Beaman, Mark 49
Bianchi, Sergio 97
Botha, Andre 16, 36, 69, 72, 73, 76, 82, 129, 130, 186, 260, 266–7, 270, 271, 277
Borrow, Nik 4, 40, 63, 66, 94, 143, 146, 155, 167, 176, 179, 193, 233, 246, 249
Booysen, Maans 7, 21, 43, 57, 60, 82, 125, 126, 190, 201, 208, 239, 255, 256
Braine, Dayne 19, 120–1, 161, 190, 259
Braine, Sean 170
Chittenden, Hugh 60, 129, 130, 204–5
Cohen, Callan 79, 88, 143
Daniels, Dick 140
Dereliev, Sergey 57, 134, 236
Drewitt, Allan 31
Doppagne, Stéphane 280
Dorse, Cliff 76, 125, 176, 190
Dowd, Tony 170
Eggert, Ron 134
Elizalde, David 85
Francis, Ian 29
Gaglio, Davide 36, 200, 201
Gedeon, Kai 149
Gibbs, Dalton 137
Ginn, Peter 215
González, Guilierrez 289
Grant, Peter 212
Hansen, Louis 103
Hills, Peter 46, 167, 230, 246
Hoddinott, David 32, 164
Hoodless, Andrew 289
Hyett, Marvin 49, 51, 140
Jasper, Louise 109, 286
Kean, Christopher 40
Kelly, Aidan 167
Krishnappa, Yathin 196
Lagerqvist, Markus 109, 274, 283
Leventis, Tasso 32, 35, 36, 43, 46, 63, 94, 100, 113, 126, 133, 155, 173, 179, 182, 212, 229, 230, 236, 242, 252, 255, 274, 280, 283
Little, Ian 82, 182, 208
Little, Rob 35, 73
Lloyd, Penn 226
Marcel, Ashute 158
Mougeot, Francois 223, back cover
Odino, Martin 52-53
Oláh, János 79
Pitteloud, Jacques 60, 69, 112, 113, 126, 134, 152, 167, 182, 239, 245, 277
Poliza, Michael 23, 196
Portier, Bruno 280
Prophet, Matt 218–9
Pugh, Andy 94, 176
Redman, Nigel 88
Riley, Adam 63, 137, 193
Rovero, Francesco 106
Shapiro, Dubi 263
Shiff, Yael 100
Shirihai, Hadoram front cover, 158
Schmidt, Otto 88, 146, 263
Skov, Bjarne 164
Sloan, Margaret 286
Steyn, Peter 36
Stockenstroom, Eric 196, 256
Suter, Werner 109
Tizard, Robert 116
Vorster, Eugene 226
Walton, Jessie 35, 72, 185, 200
White, Ian 8, 12, 18, 22, 57, 76, 125, 126, 129, 130, 185, 186, 197, 201, 226, 255, 256, 259, 260, 270, 277
Wilby, Derrick 215

AUTHOR'S NOTE

(Photo by Maans Booysen)

Rob Little was born in Durban, KwaZulu-Natal, schooled at King Edward VII High School and his career in conservation began as a Forester in 1974. He attained national Diplomas in Forestry and Nature Conservation, a BSc degree in Wildlife Resources from the University of Idaho, USA, and a PhD degree on the behavioural ecology, management and utilisation of the Grey-winged Francolin from the University of Cape Town. During 1974-1985, he was a forestry researcher, lecturer in forestry and nature conservation and warden of the Cathedral Peak mountain catchment reserve in the KwaZulu-Natal Drakensberg. During 1988-1997, Rob co-ordinated the Gamebird Research Programme at the Percy FitzPatrick Institute of African Ornithology, University of Cape Town. Together with Tim Crowe, he published the book *Gamebirds of Southern Africa*. Rob was Director: Conservation at WWF South Africa during 1997 until 2008. In April 2009 he was appointed Manager of the Centre of Excellence (CoE) at the Fitztitute. He is the Fitztitute's link with the National Research Foundation and the Department of Science & Technology and manages the research activities which are funded by the CoE using Birds as Keys to Biodiversity Conservation.

Rob has published over 70 papers in peer-reviewed journals, including book reviews, over 90 semi-popular articles, 4 books and 7 book chapters, including 14 species accounts in *The atlas of southern African birds* (Vol. 1, 1997), 12 species accounts in *Roberts' Birds of Southern Africa* (Vol. VII, 2005), and 25 species texts in *The Ultimate Companion for Birding in Southern Africa* (Vol. 1, 2014). Rob also Guest Edited a special issue of *Ostrich* (2015, Vol. 86(1&2)) in memory of Phil Hockey which attracted 20 papers, 17 with Fitztitute authors and 12 with Phil as co-author.

CONTENTS

FOREWORD 13

AFRICA: POLITICAL MAP 15

INTRODUCTION 16

GUINEAFOWLS & CONGO PEAFOWL 25
    White-breasted Guineafowl 29
    Black Guineafowl 32
    Helmeted Guineafowl 35
    Plumed Guineafowl 40
    Crested Guineafowl 43
    Vulturine Guineafowl 46
    Congo Peafowl 49

FRANCOLINS & PARTRIDGES 53
    Crested Francolin 57
    Coqui Francolin 60
    White-throated Francolin 63
    Schlegel's Francolin 66
    Shelley's Francolin 69
    Grey-winged Francolin 72
    Orange River Francolin 76
    Archer's Francolin 79
    Red-winged Francolin 82
    Finsch's Francolin 85
    Moorland Francolin 88
    Ring-necked Francolin 91
    Forest (Latham's) Francolin 94
    Barbary Partridge 97
    Sand Partridge 100
    Udzungwa Forest Partridge 103
    Rubeho Forest Partridge 106

OPPOSITE An African Snipe that has just probed its bill into the soft sediment (photo by Ian White)

GAMEBIRDS

    Madagascan Partridge 109

    Stone Partridge 112

    Nahan's Partridge 116

SPURFOWLS 121

    Red-necked Spurfowl 125

    Swainson's Spurfowl 129

    Yellow-necked Spurfowl 133

    Grey-breasted Spurfowl 137

    Erckel's Spurfowl 140

    Djibouti Spurfowl 143

    Chestnut-naped Spurfowl 146

    Black-fronted Spurfowl 149

    Jackson's Spurfowl 152

    Handsome Spurfowl 155

    Mount Cameroon Spurfowl 158

    Swierstra's Spurfowl 161

    Ahanta Spurfowl 164

    Scaly Spurfowl 167

    Grey-striped Spurfowl 170

    Double-spurred Spurfowl 173

    Hueglin's Spurfowl 176

    Clapperton's Spurfowl 179

    Hildebrandt's Spurfowl 182

    Natal Spurfowl 185

    Hartlaub's Spurfowl 190

    Harwood's Spurfowl 193

    Red-billed Spurfowl 196

    Cape Spurfowl 200

QUAILS 206

    Common Quail 208

    Harlequin Quail 212

    Blue Quail 215

SANDGROUSE 219

- Pin-tailed Sandgrouse 223
- Namaqua Sandgrouse 226
- Chestnut-bellied Sandgrouse 229
- Spotted Sandgrouse 233
- Black-bellied Sandgrouse 236
- Yellow-throated Sandgrouse 239
- Crowned Sandgrouse 242
- Black-faced Sandgrouse 245
- Lichtenstein's Sandgrouse 249
- Four-banded Sandgrouse 252
- Double-banded Sandgrouse 255
- Burchell's Sandgrouse 259
- Madagascan Sandgrouse 263

SNIPES & EURASIAN WOODCOCK 267

- Greater Painted-snipe 270
- Jack Snipe 274
- African Snipe 277
- Great Snipe 280
- Common Snipe 283
- Madagascan Snipe 286
- Eurasian Woodcock 289

GLOSSARY 292

BIBLIOGRAPHY AND FURTHER READING 297

INDEXES *(Common English, French & scientific names)* 298

FOREWORD

Africa is well known for its rich and diverse gamebird fauna, a diversity highlighted by this much-needed monograph on the terrestrial gamebirds and snipes of Africa. This important and beautifully illustrated book will without doubt fill an important gap in our literature and will be prized by ornithologists, birders, wing-shooters, land owners and conservation-minded people throughout Africa and across the world. It contains a fascinating, authoritative compilation of evolutionary, behavioural and ecological information which provides insights into the lives of these unique and special birds. Of vital importance is the fact that this information has huge significance for the future conservation of terrestrial gamebird species in Africa.

This book collates information on gamebird identification, habitat use, habits, foraging and breeding behaviour, status and conservation of each species in an informative and readable content. Illustrative photographic plates convey the overall appearance, diagnostic features, behavioural activities and in some cases the preferred habitats of each species. The information is scientifically sound and represents a unique central repository of information on Africa's gamebirds, which makes it an excellent reference source. The 74 species of upland gamebirds and snipes found in Africa fall into six groups: guineafowls and the Congo Peafowl (7 species), francolins and partridges (20 species), spurfowls (24 species), quails (3 species), sandgrouse (13 species), and snipes and the Eurasian Woodcock (7 species). *Terrestrial Gamebirds & Snipes of Africa* offers a concise summary of the large but scattered body of accumulated scientific research and field-guide literature. This book will be a worthy addition to the ornithological literature and to the bookshelves of bird enthusiasts.

The author, Rob Little, is no stranger to gamebird research and conservation. Rob coordinated the Gamebird Research Programme at the Percy FitzPatrick Institute of African Ornithology, University of Cape Town. Together with Tim Crowe, he published the book *Gamebirds of Southern Africa*. Rob was Director: Conservation at WWF South Africa from 1997 until 2008. In 2009 he was appointed Manager of the DST-NRF Centre of Excellence at the Fitztitute. He has an obvious infectious passion for this fascinating group of birds, and provides wonderful insights into their life histories. This superbly written book holistically captures the beauty and character of each species.

OPPOSITE A male Burchell's Sandgrouse takes off from a watering hole (photo by Ian White)

I believe this comprehensive and informative book will continue to stimulate an appreciation and interest in Africa's gamebirds, and will ensure increased protection of their habitats. To the reader, this book will provide many hours of informative and knowledge-enriching pleasure, providing key insights into the lives of these wonderful birds.

Nicky Oppenheimer
24 February 2016

AFRICA: POLITICAL MAP

INTRODUCTION

In *Terrestrial Gamebirds & Snipes of Africa*, I summarise and present useful and pertinent facts about these birds in an informative and readable fashion. Numerous photographs accompany the descriptive text and, where possible, reflect the appearance, characteristics, sexual dimorphism features and special behavioural activities or favoured habitat features to assist with the identification of the species. The book covers the 74 species of upland gamebirds and snipes found in Africa which comprise six groups: guineafowls and Congo Peafowl (7 species), francolins and partridges (20 species), spurfowls (24 species), quails (3 species), sandgrouse (13 species) and snipes and Eurasian Woodcock (7 species). Three endemic Madagascan species (a partridge, a sandgrouse and a snipe) and four African species (introduced Helmeted Guineafowl, Harlequin Quail, Common Quail and Greater Painted-snipe), which are also found in Madagascar, are included for regional completeness.

Gamebirds have been utilised by humans for food and sport for centuries because they are believed to withstand 'sustainable harvesting' and have also been domesticated for food production and for ornamental aviculture. They are a familiar, yet often under-appreciated, group of birds, members of which can be seen along road sides, open farmlands, in urban fringes and even within cities, towns and gardens. It is therefore not surprising that the Helmeted Guineafowl *Numida meleagris* appears more regularly as African curios on mantelpieces in living rooms and on printed African textiles than any other bird.

Helmeted Guineafowls are the best known African gamebird and are icons of the African landscape (photo by Andre Botha).

All gamebirds can be divided broadly by habitat into two artificial groups – landfowl and waterfowl – which today have representatives on all continents except Antarctica. Landfowl and waterfowl together occupy the second oldest branch of the avian evolutionary tree, after ostrich-like birds. The landfowl, including pheasants, quails, grouse, partridges, francolins, spurfowls, guineafowls and turkeys, occupy virtually all terrestrial habitats ranging from lowland deserts and tropical rainforests to Siberian tundras and the snowy slopes of the highest mountains of the Himalayas. Modern waterfowl, such as ducks, geese and snipes, are largely dependent on aquatic habitats.

In the strict sense of the word, most landfowl are chicken-like species in the avian taxonomic order of Galliformes. Traditionally, Galliformes are divided into nine major groups: megapodes, cracids (guans, chachalacas and curassows), guineafowls, grouse, turkeys, pheasants, partridges (including francolins and spurfowls), Old World quails and New World quails. Megapodes and cracids are the most ancient galliforms and are confined respectively to the tropical and subtropical regions of Australasia and South America. Megapodes have, as their name implies, large feet and differ from all other birds in that their eggs are incubated artificially, such as by heat from rotting vegetation piled in a compost-heap-like nest or from sand heated by volcanic activity. Cracids differ from the other galliforms in that they spend relatively little time scratching on the ground for food, preferring rather to pluck fruits and flowers from the branches of trees or to pick up fallen items from the ground.

Investigation of the genetics and anatomy of living species and fossils, as well as plumage, calls and behaviour, suggests that the major evolutionary break in the galliform evolutionary tree is the African guineafowls. Thus, the primary breakup of the megacontinent Gondwana into Australasia, South America and Africa more than 90 million years ago led to the creation of the three most ancient groups of galliforms on now well-separated continents. The next major event probably dates back to about 55 million years ago when partridge/francolin-like galliforms dispersed fairly simultaneously from Africa to North America and Asia respectively, giving rise to the rest of the Earth's gamebirds: quails, partridges, peafowls, junglefowls, pheasants, turkeys and grouse. The American quail descendants are relictually represented in Africa by the evolutionarily enigmatic Stone Partridge *Ptilopachus petrosus* from the African savannas north of the tropics and Nahan's Partridge *P. nahani* from the forests of the Democratic Republic of the Congo and Uganda, respectively. The Eurasian landfowl are descended from the Udzungwa Forest Partridge *Xenoperdix udzungwensis* discovered during the early 1990s from forests confined to the tops of the Udzungwa Mountains in Tanzania.

Evidence shows that pheasants, partridges, quails and francolins as understood throughout the 20th century are not natural evolutionary groups but rather mosaics of evolutionarily unrelated species. Even the best-known gamebird, the Junglefowl *Gallus gallus* (which is the wild ancestor of the domestic chicken), has shifted its 'perch' on the gamebird evolutionary tree from that of the pheasants to one including a mix of 'francolins' and 'partridges'. To make matters more confusing, 'the' Partridge *Perdix*

perdix links with the true pheasants (for example, *Phasianus* species with which it shares more than 14 tail feathers), which, in turn, do not include the peafowl *Pavo* species, the peacock pheasants *Polyplectron* species and the argus 'pheasants' (*Rheinartia* and *Argus* species), which are in fact related to one another. Furthermore, the most evolutionary artificial groups of all are the partridges and 'francolins', whose members appear to link variously with the New World quails, the Old World quails and the peacock pheasants.

Juvenile Natal Spurfowls showing the typical broad head stripes of spurfowl chicks (photo by Ian White)

The distinction between the partridge-like francolins ('true' francolins related to Asiatic *Francolinus* species) and the pheasant-like francolins (spurfowls) is apparent in the plumage of their downy chicks. They differ in facial markings and the markings on their backs and crowns. Francolins have quail-like back feathers with strong barring and a streak along the quill shaft. The back feathers of spurfowls have no shaft streak and are, at best, extremely irregularly barred. The crowns of francolin chicks are multi-striped as opposed to the broadly striped spurfowl chicks.

From a conservation perspective, research on gamebirds is essential because, time and again, the one biological 'bond' that unites gamebirds (their ability to be 'harvested' sustainably) has been broken. In particular, information on gamebird behavioural ecology and demographics can highlight the roles of changes to the natural environment by humans that can, on the one hand, produce some of the most spectacular wing-shooting in the world, but can also destroy it and harm biodiversity across the board. The bottom line is that relatively easy-to-apply management of the landscape can be a 'win-win' conservation success, whereas mis-management can result in disaster. Unfortunately, it is safe to say that much of Africa has been subject to extensive habitat transformation and degradation, mostly through conversion to croplands, extensive frequent burning, as well as to urban areas and rural settlement. I trust that the information provided on the human threats to gamebirds will stimulate effective implementation of practical conservation guidelines that restore and maintain gamebird habitats and populations.

The aim of this book

I aim to collate and summarise the large, but scattered, body of literature on the biology of African gamebirds and snipes, and to present that information in this book. Much of the information contained is gleaned from the *Handbook of the Birds of the World* (Del Hoyo et al. 1994, 2013), *The Birds of Africa Volume II* (Urban et al. 1986), *Birds of Africa South of the Sahara* (Sinclair & Ryan 2010) and *Gamebirds of Southern Africa* (Little & Crowe 2011). The species texts present information on the classification, description, distribution and habitat, habits, food and feeding, breeding, status and conservation of each species in a readable format. By doing this, I hope to encourage birders, conservationists, landowners and wing-shooters to appreciate this special group of birds and to manage gamebirds as 'umbrella' species, whose conservation promotes biodiversity conservation in general.

This Orange River Francolin is calling from a conspicuous perch to advertise his presence in his territory (photo by Dayne Braine).

How to use and interpret this book

The main content of the book is contained within six sections, which represent the overall groups covered by the book. Each group section has an introduction that briefly highlights the diagnostic features of that group, their natural history and the evolutionary relationships between the members of the group.

Each species account opens with the currently accepted English and French common names, followed by the current scientific name of the bird. The third line of the opening text refers to the original scientific name and details of its discovery, including the surname of the scientist who was the first to describe the species, the year in which this was done, and the place of the first record. The naming of the gamebirds and snipes of Africa largely follows the International Ornithological Congress (IOC) World Bird List (version 5.3, Gill & Donsker 2015).

The distribution map for each species reflects the current distribution of the species within the continent. Due to the scale of the maps, only relatively vague interpretation of distribution can be ascertained at a local scale.

Classification

This section gives the most current information on the species' evolutionary status and affinities. The most closely related species are given where appropriate. Where the species varies significantly geographically, distinctive forms are recognised as subspecies, which are listed, along with their general geographical distribution. This information largely follows the IOC World Bird List (version 5.3, Gill & Donsker 2015), Tim Crowe's overview of gamebird evolution (The Conversation, 24 February 2016) and the PhD dissertation of Tshifhiwa Gift Mandiwana-Neudani (2013, University of Cape Town).

Description

This section will help the reader to identify the species in the field and to distinguish adults, juveniles and chicks using, where available, information on overall form and plumage. Some comments are also given to differentiate from similar species. Identification can also be assisted by referring to the photographs, which, where possible, reflect the general appearance, characteristic and sexual dimorphism features of each species.

Distribution and habitat

Here I reflect the species' geographical range and their habitat preferences. Information should be interpreted in conjunction with the distribution map of each species.

Habits

This part summarises what is known about the species' behavioural ecology, any movement patterns, and their displays and calls, both day by day and throughout the year.

Food and feeding

This section describes any specific foraging behaviour and identifies food types taken on a seasonal basis.

A male Crested Guineafowl offering a courtship food item to a hen while she dust-baths (photo by Maans Booysen)

Breeding

This section gives information on the species' mating system (for example, if it is monogamous), breeding (egg-laying) and factors (such as rainfall) that may influence its timing, the architecture of its nest, clutch size, egg colour and size, incubation period and the growth and development of chicks and juveniles.

Status and conservation

This part indicates the status of the species' populations evaluated on a geographical basis, emphasising any potential threats of local populations and where and if (and when) it may be utilised sustainably by wing-shooters.

GAMEBIRDS

The glossary

Because the species accounts contain biological information, terms may be used with which some readers are not familiar. These are explained or defined in the glossary.

Bibliography and suggested reading

The bibliography at the end of the book presents a brief list of key references on which I relied for information. While the bibliography is not fully comprehensive, it is sufficiently detailed to allow any reader to source the key studies of the 74 species represented in the book. Further reference material can be found in the reference sections of the *Handbook of the Birds of the World* (Del Hoyo et al. 1994, 2013), *The Birds of Africa Volume II* (Urban et al. 1986), *Birds of Africa South of the Sahara* (Sinclair & Ryan 2010) and *Gamebirds of Southern Africa* (Little & Crowe 2011).

Sandgrouse, in this case a male Burchell's Sandgrouse, belly-wet to transport water to their flightless chicks in harsh arid environments (photo by Ian White).

OPPOSITE A group of Red-billed Spurfowls catching the early morning sun (photo by Michael Poliza)

PART 1
Guineafowls & Congo Peafowl

GUINEAFOWLS AND CONGO PEAFOWL

Classification

Guineafowls make up the Afrotropical endemic family Numididae, with six species in four genera:

Agelastes Bonaparte, 1850, containing two species:

- White-breasted Guineafowl *A. meleagrides* of West Africa
- Black Guineafowl *A. niger* of Central Africa

Numida Linnaeus, 1764, containing one species:

- Helmeted Guineafowl *N. meleagris* are characteristic of virtually all open-country habitats outside deserts and alpine systems throughout sub-Saharan Africa.

Guttera Wagler, 1832, containing two species:

- Plumed Guineafowl *G. plumifera* are confined largely to the primary rainforests of central equatorial Africa.
- Crested Guineafowl *G. pucherani* of western, central and southern Africa, which inhabit forest edges, gallery forest along rivers, and secondary forest regrowth.

Acryllium Gray, GR, 1840, containing one species:

- Vulturine Guineafowl *A. vulturinum* extends the realm of the guineafowl into the semi-deserts of eastern Kenya, southern Ethiopia and Somalia.

Recent research suggests that the Kenya Crested Guineafowl *G. edouardi* of East Africa may be recognised as a full species in the future. Although the Congo Peafowl *Afropavo congensis* is more closely related to Asian peafowls, *Pavo* species, and the pheasants in the family Phasianidae than to the guineafowls, they are placed with the guineafowls here because they may have more of a field affiliation with them in Africa.

Description

- Guineafowl species differ from one another primarily in the colour of the naked skin on their head and eye surrounds, in the colour of other facial structures and adornments, and in the pitch of their advertisement cackles. These anatomical and behavioural characteristics are functional components within the individual species' mating systems, allowing them to instinctively distinguish potential sexual partners of their own species from members of the other guineafowl species. There is sometimes striking, effectively qualitative, geographical variation in head colour

PREVIOUS PAGE Crested Guineafowl subspecies *Guttera pucherani pucherani* (photo by Lorenzo Barelli) and adornments among various subspecies (or geographical races). However, none of this variation has any demonstrable adaptive value, plays any role in mate recognition or inhibits interbreeding between members of the various subspecies. For example, when domesticated Helmeted Guineafowl, previously derived from the West African subspecies, is bred in captivity with members of other subspecies from as far afield as Somalia and East Africa, or is released into populations of wild Helmeted Guineafowl in southern Africa, it interbreeds freely without any apparent loss of fertility. Indeed, within the zones of natural geographical contact between subspecies of the Helmeted Guineafowl, in particular, there is apparently free interbreeding, producing anatomically intermediate forms of all persuasions.

- The other prominent feature that distinguishes one guineafowl 'type' from another is the pitch of the advertisement calls, which seems to be an adaptation that allows effective communication in different habitats. For example, the Black Guineafowl, which lives in dense primary rainforest, has a low-pitched, almost musical call designed to penetrate thick vegetation. The calls of the Crested Guineafowl, which lives in less dense, secondary forest and riverine bush, are somewhat higher pitched. The call of the Helmeted Guineafowl, which lives in open savanna habitat, is higher pitched still. Finally, the Vulturine Guineafowl *Acryllium vulturinum*, which inhabits more open, semi-arid scrubby grassland, has the highest-pitched call of all.

Natural history

Guineafowls occur in virtually all African habitats from dense rain forest to semi-desert, only excluded from extreme deserts and extensive wetland landscapes. They are generally gregarious, forming flocks during the non-breeding season and flocks can even converge in areas of optimal foraging. They are icons of the African landscape and are depicted on more African curios than any other bird on the continent.

They tend to have large clutches and chicks, commonly known as keets, remain with the parents into the non-breeding season to form flocks. Guineafowl are favoured for food by many rural communities and as such are captured or even domesticated, particularly the Helmeted Guineafowl, to breed for the table.

Evolutionary placement

- The evolution of guineafowls, like that of the Australasian megapodes (Megapodiidae) and the largely South American cracids (Cracidae), was influenced by the break-up of Gondwana (the southern mega-continent comprising Australia, India, Africa and South America) more than 100 million years ago.
- Guineafowls bridge the evolutionary gap between the relatively advanced, pheasant-like birds and the anatomically relatively primitive megapodes and cracids.
- Therefore, Africa is the likely origin of a guineafowl-like gamebird that was the ancestor of modern galliform gamebirds.

- Evolution has produced, in some instances, new species and genera of guineafowl reproductively isolated from one another and, in others, merely geographical variants that interbreed freely.
- The Congo Peafowl is endemic to the Congo Basin, and is one of three extant species of peafowl, the other two being the Indian Peafowl *Pavo cristatus* originally from India and Sri Lanka, and the Green Peafowl *P. muticus* native to Burma and Indochina.
- The Congo Peafowl was only described as a species in 1936 by Dr James Chapin who noticed that local Congolese headdresses contained long reddish-brown feathers that were not consistent with any known central African bird. He later found two specimens in the Royal Museum of Central Africa with similar feathers labelled as the 'Indian Peacock' which he discovered to be the Congo Peafowl, a completely different species. In 1955 he located seven live specimens of the species.
- The Congo Peafowl has physical characteristics of both peafowls and guineafowls which may indicate that the Congo Peafowl is a link between the two families.

Conservation

Five of the six guineafowl species are relatively secure with only the White-breasted Guineafowl being severely threatened, and thus listed as Vulnerable (The IUCN Red List of Threatened Species, version 2015-4). While all six species are impacted by hunting pressure, including shooting and snaring, at various levels across their range, the forest species are most impacted by habitat fragmentation and loss due to deforestation and the more open habitat species are influenced by agriculture, including pesticides and the injudicious use of fire.

White-breasted Guineafowl
Pintade à poitrine blanche
Agelastes meleagrides

Agelastes meleagrides 'Temm.' Bonaparte, 1850, the type locality is probably Ghana

White-breasted Guineafowls photographed in Tai Forest, Ivory Coast
(both photos by Ian Francis)

Classification

The White-breasted Guineafowl is most closely related to the Black Guineafowl *A. niger*. They are monotypic with no recognised subspecies.

Description

- They are easily identified by their completely bare pink or red head and neck, conspicuous white neck collar and breast, and otherwise black body.
- Their bill is greenish brown and their legs are greyish black.
- Juveniles have no white collar, the head and upper neck are covered with brownish-black downy feathers, with two tawny crown stripes, and their belly is white.
- Downy keets are greyish brown with pale ochre stripes on a dark sepia head and neck.

Distribution and habitat

- They are uncommon to rare from south-eastern Sierra Leone, Liberia, Ivory Coast to western Ghana.
- See status and conservation for indications of their local extinction.
- They inhabit primary West African lowland rainforests, which have an open or sparse undergrowth, or older regrowth adjacent to primary forest.

Habits

They occur in loose groups of up to 20 birds, and are shy and difficult to approach. Although they rarely leave the protective forest cover, they do not favour dense undergrowth. They roost in bushy trees in the understory. They utter a metallic twitter when foraging, a loud, even-pitched whistle to muster the flock, and a descending, flute-like whistle by isolated individuals.

Food and feeding

They forage on the ground below fruiting trees, scratching among soil and leaf litter with their feet like domestic chickens. Their diet consists mainly of berries and fallen seeds of forest trees, and insects including termites, ants, crickets, beetle larvae, as well as millipedes, spiders and small molluscs.

Breeding

Not much is known about their breeding biology. Breeding is during October to May, peaking from November to January. They nest on the ground in dense undergrowth. The clutch of up to 12 eggs are reddish buff with white pores and measure 45.0 x 35.0 mm.

Status and conservation

They are severely threatened by hunting pressure and habitat destruction and may disappear except from a few protected areas if efforts are not made to secure more representative portions of their populations. Populations in Ghana, Sierra Leone and Guinea may be largely extinct, and they are rare in Liberia and Ivory Coast. Although shy and secretive, they are easily trapped and hunters also lure them by imitating the call and killing them with a slingshot. A recent census conducted in the Tai Forest, Ivory Coast, suggests that they are still fairly numerous, despite the poaching and habitat-related issues, along with civil unrest. But their status in Ivory Coast and Liberia remains precarious in the long-term. Their conservation status is therefore listed as Vulnerable (The IUCN Red List of Threatened Species v 2015-4).

White-breasted Guineafowls in typical dense forest habitat (photo by Allan Drewitt)

Black Guineafowl
Pintade noire

Agelastes niger

Phasidus niger (Cassin, 1857), Cape Lopez, Gabon

Black Guineafowl photographed in Gabon (photo by Tasso Leventis)

Black Guineafowl photographed in Gabon (photo by David Hoddinott)

GUINEAFOWLS & CONGO PEAFOWL

Classification

The Black Guineafowl was previously placed in the monotypic genus *Phasidus*, but they are most closely related to the White-breasted Guineafowl *A. meleagrides*. There are no subspecies recognised.

Description

- They are a relatively small (average 700 g), entirely plain black guineafowl with a bare dull red head and throat and a short black downy crown and hind-neck crest.
- The rest of the plumage is unmarked black, and the wings and tail are unmarked.
- The bill is greenish grey and the legs are greyish brown.
- Their lack of plumes on a pinkish head and unspotted plumage distinguish them from the larger Plumed Guineafowl *Guttera plumifera*, which has a similar distribution.
- The sexes are similar in appearance with the male being slightly larger.
- Juveniles are dull black, have a feathered neck, dull brown head and a greyish white belly.
- Downy keets are mainly dark rufous and black above.

Distribution and habitat

- They are endemic to Central Africa from Cameroon, Equatorial Guinea, Gabon and far-northern Angola, through the central Democratic Republic of the Congo to the western edge of the Rift Valley.
- They are scarce to locally common, but seldom seen.
- They are found chiefly in dense primary lowland rain forest mostly with rank undergrowth, also in forest edges and on old, overgrown fields.

Habits

They are usually encountered as elusive pairs or small groups of less than 10 individuals, which frequent the forest floor and rapidly run within dense primary forest thickets for cover. They are presumed to roost in trees or bushes. They have a resonant rising, rather melodious '*huwhee-huwhee-huwhee-huwhe- wheet-wheet-wheet …*' whistling call, and a '*keet keet*' contact call.

Food and feeding

They forage as a flock on the forest floor, scratching with their feet. Their diet consists of hard seeds, green leaves, fruits, beetles, ants, termites, millipedes and small frogs.

Breeding

They are probably monogamous. In north-eastern Zaire, they breed in almost any month, but mainly during the drier months of December to February. The nest, clutch size and incubation period are undescribed. The deeply pitted eggs are pale, reddish brown, sometimes washed with yellow or violet and measure 42.0 x 34.0 mm.

Status and conservation

They are probably less threatened by habitat loss than White-breasted Guineafowl because their range is much more extensive, but are nonetheless vulnerable to forest destruction and hunting. The viability of protected areas within their range is probably vital to their long-term conservation.

Helmeted Guineafowl
Pintade sauvage

Numida meleagris

Phasianus meleagris (Linnaeus, 1758),
Upper Nile, Nubia, Sudan

Helmeted Guineafowl keets in the Kruger National Park, South Africa (photo by Jessie Walton)

Helmeted Guineafowl subspecies *N. m. galeatus* from Nigeria (photo by Tasso Leventis)

A one-day-old Helmeted Guineafowl chick (photo by Rob Little)

GAMEBIRDS

Helmeted Guineafowl subspecies *N. m. reichenowi* from Tanzania (photo by Tasso Leventis)

A typical Helmeted Guineafowl nest and clutch (photo by Peter Steyn)

Helmeted Guineafowl in flight (photo by Davide Gaglio)

Helmeted Guineafowl at a water hole (photo by Andre Botha)

Classification

Nine subspecies are recognised:

N. m. sabyi Hartert, 1919, between the Oum er Rbia and Sebou rivers in north-western Morocco

N. m. galeatus Pallas, 1767, in West Africa, east to southern Chad, and south to central Democratic Republic of the Congo and northern Angola, has a distinctive white face and rounded, red wattles

N. m. meleagris (Linnaeus, 1758) in eastern Chad, east to Ethiopia and south to northern DRC, Uganda and northern Kenya, has a small helmet, a blue face, rounded blue wattles and a short, cartilaginous, bristly 'moustache'

N. m. somaliensis Neumann, 1899, in north-eastern Ethiopia and Somalia, has a very long moustache and triangular wattles, which are blue at the base and red at the tips

N. m. reichenowi Ogilvie-Grant, 1894, in Kenya and central Tanzania has a huge, sabre-shaped helmet and rounded, red wattles

N. m. mitratus (Pallas, 1764) in Tanzania, eastern Mozambique, Zambia, Zimbabwe and northern Botswana

N. m. marungensis Schalow, 1884, in the southern Congo Basin, south to western Angola and east to the Zambezi Basin, and in the Luangwa Valley in Zambia, has a long helmet and pennant-shaped, blue wattles with red tips

N. m. papillosus Reichenow, 1894 [including *N. m. damarensis* (Roberts, 1917)] in the drier parts of western Botswana and Namibia

N. m. coronata Gurney, 1868, in South Africa

Description

- They have a bony casque or 'helmet' and bare fleshy wattles.
- Sexes are similar, however, males exhibit the hump-backed display, walking erect and on their toes with wings raised and held close to his body.
- Females tend to walk flat-footed and appear slouched in posture.
- Juveniles have a tawny-buff head with two blackish lateral crown stripes and no casque.
- Their plumage is drab grey-brown, barred with reddish-brown and speckled with black.
- The head retains the feathers of the downy keet almost until full grown.
- The helmet appears at about 40 days.
- Keets have light brown and buff facial stripes with a dark brown crown stripe.

- The back plumage is mottled brown and grey, with two longitudinal buffy stripes bordered with black, straddling a central broad brown stripe.

Distribution and habitat

- They are widespread in most open-country terrain, from near-desert to the edges of forests, especially in savannas mixed with cultivation.
- They are absent from some arid areas, which lack drinking water and elevated roosts.
- They have expanded their range where humans have added these missing habitat features, even telephone poles can serve as nightly roosts.
- Humans have further promoted this range expansion by translocating wild birds into new areas, including to Madagascar.
- This has also promoted the release of domesticated Helmeted Guineafowl into the wild.
- Evidence of domestication is white feathers on the wings or body, whitish instead of blue facial skin and orange not brown-black legs.

Habits

During the non-breeding season, flocks of 15–40 birds are formed, with larger aggregations being multiple flocks converging on a super-abundant resource. Flocks descend from their roost at dawn, move to a drinking site, then socialise and dust-bath. Thereafter, they move into foraging areas or mixed bush-grassland shade cover until late afternoon. A couple of hours before sunset, they may move to a watering site to socialise and drink. Just before sunset they return to their roost. They rarely fly, only if pressed or to mount their roost. They have a cackling, staccato '*kek, kek, kek, kek, kaaaaaa, ka, ka, ka, ka, kaaaaaa, ka, ka*' alarm call. Females have a two-note 'buck-wheat' contact call used during the breeding season. Flock members keep in contact by emitting a single-noted '*cheenk*' call.

Food and feeding

Their diet consists of bulbs, primarily *Cyperus* species, the stems of plants and spillage from harvested grain, but will readily shift to grass and forb seeds. While breeding, they eat invertebrates, including agricultural pests, especially grasshoppers and termites. They do not normally take growing maize plants or maize from cobs still attached to healthy plants.

Breeding

Breeding is triggered by rainfall. The female selects the nest site in tall grass at the base of a tussock or under a bush. The nest is a well-concealed scrape in the earth, lined with feathers and grass stems. The clutch of 6–20 (and sometimes more), broad, slightly pointed oval eggs are white to pale brown with fine speckling. They are hard and thick-shelled and measure 52.6 x 40.2 mm. Incubation by the female lasts 24–27 days. Both parents aggressively defend their precocial keets, which can join their parents on the roost when 2–3 weeks old.

Status and conservation

Threats include disease, predation by natural and feral carnivores, and snaring and poisoning by humans. However, the primary causes of population declines are habitat fragmentation and destruction associated with extensive crop farming. These involve large-scale planting and the elimination of weeds and insects by the use of herbicides and insecticides, which destroy nesting cover and food resources.

Plumed Guineafowl
Pintade plumifère

Guttera plumifera

Numida plumifera (Cassin, 1857), Cape Lopez, Gabon

Perched Plumed Guineafowl in Central African Republic (photo by Christopher Kean)

Plumed Guineafowls among dense forest undergrowth in Gabon (photo by Nik Borrow)

Classification

The Plumed Guineafowl is most closely related to the Crested Guineafowl *G. pucherani*. There are two subspecies recognised:

G. p. plumifera (Cassin, 1857) in southern Cameroon down to northern Angola

G. p. schubotzi Reichenow, 1912, in northern and eastern Democratic Republic of the Congo

Description

- They are slightly smaller than Crested Guineafowl, but differ in appearance by having a long erect crest of stiff black plumes from the forehead to the rear of the crown, which is not curly.
- They have a dark grey, not a blue-grey, face and neck, but also have light blue spots on a black body.
- Adults have longish black facial wattles, not red, and no red on the throat. *G. p. schubotzi* has patches of orange skin in front of the ear and on the hind neck.
- Juveniles lack spots and have faintly scaled underparts.
- Downy keets are buffy with dark longitudinal stripes.

Distribution and habitat

- They are locally common to uncommon, but rarely seen, endemic to Central Africa.
- Their distribution is from southern Cameroon, Equatorial Guinea, coastal Gabon and Congo, across northern Congo and the Democratic Republic of the Congo, through the Congo basin eastwards to the western edge of the Rift Valley.
- Their distribution is similar to that of Black Guineafowl *Agelastes niger* but probably more patchy.
- They inhabit primary lowland rainforest and sometimes mature secondary growth.
- They do not venture into cultivated clearings.

Habits

They are shy, and non-breeding birds gather in flocks of up to 50 birds. When disturbed, they scatter and run rather than fly, but if flushed will fly into the thick foliage of tall trees and remain motionless until the threat has passed. They roost together in tall trees. They have a far-carrying '*kow kow khep, kow kow khep* …' trumpeting alarm call and a sharp '*tchik tchik*' contact call.

Food and feeding

They forage together on the forest floor, scratching up large areas of leaf litter with their feet in search of food. Their diet consists of roots, seeds, fruits, leaves and invertebrates, including snails, slugs, millipedes, spiders, roaches, grasshoppers, crickets, hemipterans, beetles, termites and ants.

Breeding

They are monogamous and breed during the wetter months. The nest is a simple scrape in the ground among dry leaves, with which it is lined. The average clutch is 10 eggs, which are pale buff, with numerous darkened pits, and measure 49.0 x 38.0 mm. Incubation lasts for 23 days.

Status and conservation

They are not threatened, but their range may be decreasing and becoming fragmented because of forest and thicket habitat destruction, combined with hunting pressure.

Crested Guineafowl
Pintade de Purcheran

Guttera pucherani

Numida pucherani (Hartlaub, 1861), Zanzibar

ABOVE Crested Guineafowl subspecies *G. p. verreauxi* in Nigeria (photo by Tasso Leventis)

LEFT Crested Guineafowl subspecies *G. p. edouardi* in southern Africa (photo by Maans Booysen)

BELOW Crested Guineafowl subspecies *G. p. pucherani* in East Africa (photo by Lorenzo Barelli)

Classification

Five subspecies are recognised, although this may be reviewed in the future:

G. p. pucherani (Hartlaub, 1861) in south-western Somalia to central Tanzania, Zanzibar and Tumbatu Island, has red bare skin around the eye and genetic differences, which may warrant full species status in the future

G. p. verreauxi (Elliot, DG, 1870; includes *G. p. kathleenae* and *G. p. schoutedeni*) in Guinea Bissau, east to north-western Cameroon and through the DRC to western Kenya, and south to Angola and western Zambia

G. p. sclateri Reichenow, 1898 in the Sanaga River region of western Cameroon

G. p. barbata Ghigi, 1905 in south-eastern Tanzania, southwards to northern Mozambique and Malawi

G. p. edouardi (Hartlaub, 1867) in eastern Zambia through southern Mozambique to north-eastern South Africa, has grey skin on the nape (this is blue in the other subspecies)

The localised *G. p. symonsi* in the Karkloof Forest of KwaZulu-Natal, South Africa is not considered a valid subspecies.

Description

They have a thick crest of soft curly black feathers, crimson eyes (brown in *G. p. verreauxi*) and a pale grey 'hood' of skin on the back of their head.

- Their lower-neck and upper-breast feathers are black, giving a dark overall appearance.
- Their body and wing feathers have a bluish tinge and pale bluish-white spots, with pale whitish-brown outer secondary flight feathers, visible at rest and as a pale band in the mid-wing in flight.
- Juveniles are duller, with heavily scaled plumage.
- Keets have a dark brown forehead and crown with a lighter rufous-buff cap.
- The sides of the head are brown, with irregular, multiple buff streaking.
- The back has a broad, central, dark brown stripe flanked by two stripes on either side, the inner stripe being buff, the outer dark brown.
- The underparts are rufous-buff, lighter on the throat.

Distribution and habitat

- They are widespread but fragmented from southern West Africa across Central and East Africa to southern Africa.
- Their preferred habitats are forest edge, secondary and gallery forest, and they spend much of their time skulking in thick bush.

- Although associated with forest and woodland habitats, they are found in both moist and semi-arid sites.
- They are a species of wild habitats, seldom venturing into cultivated farmlands.

Habits

They are relatively secretive and wary, although can be fairly tame in the vicinity of bush camps. They are sedentary and roost at night in trees. They are gregarious during the non-breeding season, with flocks averaging fewer than 20 birds. During the early morning, the flock will move into forest glades to preen and socialise in the sun. Before sunrise and after sunset, they will venture onto bush tracks where they are vulnerable to collisions with motor vehicles. When disturbed they cock their tail and scurry around in a seemingly confused manner. The alarm call is a staccato '*chuk-chuk-chukchukerr*', which has a low pitch. They are noisy, often heard before they are seen and sometimes call well into the night and during the pre-dawn hours. Members of the flock keep in contact by emitting a low-pitched '*chuk*' call.

Food and feeding

They forage mainly on the ground, scratching in leaf litter with their feet, but will also fly into trees to pick fruit and berries. They feed on the seeds, fruits and berries of Rubiaccaee, Amaranthaccae, Compositae, Malvaceae and Fabaceae. They also eat soft shoots, stems, green leaves, bulbs and roots, but are not attracted to maize scattered on the ground. Animal food includes spiders, beetles, grasshoppers, termites and millipedes. They often scavenge scraps of fruit and seed-bearing faeces dropped by monkeys from the trees above, where both benefit by reacting to each other's predator-alarm calls.

Breeding

They are monogamous and the male courtship-feeds his mate. Peak breeding activity is largely linked to rainfall. The nest is a shallow scrape in the ground in thick cover, often situated near a log, perhaps to shield it from the view of terrestrial predators. The clutch of four to eight deep buff to pinkish eggs measure 52.4 x 41.3 mm, and are less pointed than those of Helmeted Guineafowl *Numida meleagris*. Incubation by the female lasts 23 days. Keets can fly at 14 days and moult into juvenile plumage at about a month.

Status and conservation

There is no evidence of substantial expansion or contraction in their range. Although nowhere severely threatened, local range decreases might be due to destruction of forest and thicket habitats. They are not a favoured wing-shooting species and do not appear to be significantly affected by hunting pressure.

Vulturine Guineafowl
Pintade vulturine
Acryllium vulturinum

Numida vulturina (Hardwicke, 1834), Tsavo, Kenya

A group of Vulturine Guineafowls from Ethiopia (photo by Tasso Leventis)

Perched Vulturine Guineafowls in Kenya (photo by Peter W. Hills)

Classification

The Vulturine Guineafowl is monotypic, being the only species in the genus *Acryllium* Gray, GR, 1840. There are no subspecies recognised.

Description

- They are a distinctive and attractive tall slender guineafowl with long legs, a long thin neck and a long pointed tail.
- The bare head and neck are bluish grey with a patch of short dense chestnut down on the nape.
- The breast and back are bright cobalt-blue, overlaid with black-and-white lanceolate feathers.
- The secondaries are edged with lilac.
- The rest of the wings and the remainder of the body are black, with white spots.
- Their bill is light grey and their legs are black.
- Juveniles are duller greyish-brown, with black, brown and rufous bars and spots.
- Downy keets are yellowish buff, mottled with dark brown, with the centre of the crown dark brown, a dark curving streak below the eye and two curved cheek stripes.

Distribution and habitat

- They are locally common from southern Ethiopia, north-western and southern Somalia, and semi-arid eastern Kenya to far north-eastern Tanzania.
- They prefer arid, open scrub and dry savanna, but will also inhabit montane forest patches and tall riverine woodlands.

Habits

They occur in flocks of 20–30 individuals outside of the breeding season. They roost in tall trees and retire to thick bush in the heat of the day. When disturbed the flock will cluster in a group and walk or run to nearby cover. If flushed, they will fly 50–100 m and then run, seldom landing in a tree. The characteristic call is a high-pitched, metallic '*chink-chink-chink-cheenk-cheenk krrrrrrrr*' rattle, which is faster and higher pitched than that of Helmeted Guineafowl *Numida meleagris*.

Food and feeding

They forage mostly on the ground scratching with their feet, but will climb into bushes to feed on berries and fruits. Their diet consists of roots, bulbs, seeds and leaves of grasses and herbs, berries and fruit (*Commiphora* spp. and *Ficus* spp.), green buds and shoots. They also eat insects, scorpions, spiders and small molluscs. They are apparently independent of fresh water, and do not drink even where water is readily available in the dry season.

Breeding

The timing of breeding is triggered by rainfall, peaking in June and December/January. The nest is a simple scrape in the ground in thick grass or bush cover, sparsely lined with grass. The clutch of 4–8 creamy white or pale brown, smooth and slightly pitted, broad oval and slightly pointed eggs measure 50.0 x 36.0 mm. Incubation is for 23–25 days. In general, information on their breeding biology in the wild is lacking.

Status and conservation

They are generally not threatened, except by local hunting pressure.

Congo Peafowl
Paon du Congo

Afropavo congensis

Afropavo congensis Chapin, 1936, Sankuru district, central Democratic Republic of the Congo

ABOVE A male Congo Peafowl photographed in the Bronx Zoo, New York, USA (photo by Marvin Hyett)

LEFT A roosting female Congo Peafowl from the Lomako Forest, Democratic Republic of Congo (photo by Mark Beaman)

Classification

The Congo Peafowl is more closely related to the Asian peafowls, *Pavo* species, and the pheasants in the family Phasianidae than to the guineafowls in the family Numididae. They are monotypic with no subspecies recognised.

Description

- They are larger (ca 1 400 g) and much more brightly coloured than the Black Guineafowl *Agelastes niger* and the Crested Guineafowl *Guttera puncherani*.
- Males are deep iridescent violet-green, with a red throat, blue breast and bristled black-and-white crown tuft.
- Females are duller than the males, with a greenish back, a rufous head tuft and underparts, and green-barred rufous wings.
- Juveniles have duller upperparts with a brown-tinged back and dull black underparts.
- Downy chicks have a black crown, nape and hind neck, tawny yellow sides of the head with a narrow black line along the eyelid and above the eye, and a small black patch just behind the eye and on the ear covert.

Distribution and habitat

- They are endemic to the equatorial forests of central and eastern Democratic Republic of the Congo and their overall biology is poorly known.
- They may be locally common in remote areas, favouring undisturbed primary rainforest below 1 200 m above sea level with extensive undergrowth on well-drained soils.
- They are known to occur in the Maiko, Salonga and Ituri Forest National Parks.
- They avoid seasonally flooded areas.

Habits

They occur in pairs or small groups and are very secretive within the dense understory. They are mostly crepuscular, retreating to hide in undergrowth during most of the day. They can also be active well into the night, particularly when the moon is full. They roost high in trees at night. When disturbed they will cock and spread their tail and erect their crest. The call is a repeated duet '*goway-gowaa*' with the male calling first, and is higher-pitched than the female.

Food and feeding

They forage on the forest floor among the dense undergrowth. Their preferred diet includes the seeds and fruits of trees and forest understory plants, including *Celtis adolfi-friedericii*, *C. ituriensis*, *Sapium ellipticum*, *Ficus* spp., *Strombosia grandifolia*, *Xymalos monospora* and *Caesalpinia decapetala*, as well as ants, termites and their larvae, and other terrestrial and aquatic invertebrates.

Breeding

They are monogamous, aggressive and males display standing erect with their tail and wings spread, similar to other peafowls *Pavo* spp. and turkeys *Meleagris* spp. They also display sideways like other phasianids. The eggs are laid in a flat spot in a branched fork of a tree. The clutch of 2–4 eggs are plain rufous-brown or cream and measure 60.0 x 45.0 mm. Incubation by the female lasts for 25–28 days. The chicks are brooded by the female in the nest for two days, whereafter they descend to the ground.

Status and conservation

Their populations may be impacted by hunting and they are susceptible to disease and parasites. As they are also restricted to secluded forest areas, forest destruction will be a major limiting factor. Their conservation status is therefore listed as Vulnerable (The IUCN Red List of Threatened Species v 2015-4).

A male Congo Peafowl photographed in the Bronx Zoo, New York, USA (photo by Marvin Hyett)

PART 2
Francolins & Partridges

FRANCOLINS & PARTRIDGES

Classification

- Twelve of the 13 African francolin species of the genera *Afrocolinus* (gen. nov. Mandiwana-Neudani 2013), *Dendroperdix* Roberts, 1922, *Peliperdix* Bonaparte, 1856, and *Scleroptila* Blyth, 1849, within the family Phasianidae, can be grouped into three groups according to diagnostic plumage features: striated, red-winged and red-tailed (see the Table 1).
- The Forest Francolin *Afrocolinus lathami* represents its own evolutionary trajectory and is not included in the francolin groups.
- Furthermore, the Crested Francolin will likely be reassigned from the genus *Dendroperdix* to the genus *Ortygornis* Reichenbach, 1852.

Table 1: Recognised francolin species groups (*Dendroperdix*, *Scleroptila* & *Peliperdix* ssp.)

Striated	Red-tailed	Red-winged
Crested *Dendroperdix sephaena*	Coqui *Peliperdix coqui*	Shelley's *Scleroptila shelleyi*
	White-throated *P. albogularis*	Grey-winged *S. afra*
	Schlegel's *P. schlegelii*	Orange River *S. levalliantoides*
		Archer's *S. gutturalis*
		Red-winged *S. levaillantii*
		Finsch's *S. finchi*
		Moorland *S. psilolaema*
		Ring-necked *S. streptophora*

Description

- Francolins are relatively small, Old World, partridge/quail-like terrestrial gamebirds, weighing 220–550 g.
- They have quail-like back feathers (chestnut, barred with black and white, and with a white shaft streak).

PREVIOUS PAGE A male Coqui Francolin subspecies *P. c. hubbardi* from Lake Nakuru National Park, Kenya (photo by Martin Odino)

- Their legs are generally yellow and males have only one short, relatively blunt spur situated about half-way down the lower leg.

Natural history

Francolins tend to be sedentary and occur in grassland habitats, grading towards those with scattered bush, to a few inhabiting extensive patches of thicket. They sit tight on the ground when disturbed and at night, eventually flushing as a group with a whir of wings, dropping back to the ground after the flush-flight. Their calls are typically musical and whistling, and they react strongly to calls played back to them. The incubating hen sits very tight, particularly late in the incubation period. Their chicks have a narrow, dark crown patch on their heads, bordered by several black and white stripes above and around the eye. Chicks are precocial, leaving the nest about two hours after hatching, and are cared for by both parents. If a predator approaches, the parent closest to the chicks will attempt to distract the predator by limping away, feigning injury.

Evolutionary placement

- Traditionally, francolins (and spurfowls) have been classified with partridges in the tribe Perdicini within the family Phasianidae, although the only anatomical feature that supports this grouping unambiguously is that they both have 14 tail feathers.
- Until recently, all 41 species of francolins and spurfowls (36 African and 5 Asiatic) were lumped into a single genus, *Francolinus*, but divided among seven groups, with at least one member of which may be found in all habitats throughout sub-Saharan Africa outside of true desert.
- Many birders, farmers and wing-shooters divide these groups into two assemblages:
    - partridges (which make up the striated, red-winged and red-tailed francolin groups); and
    - francolins (of the bare-throated, montane, scaly and vermiculated groups).
- Anatomical and genetic research suggests that francolins and spurfowls are indeed natural, but distantly related evolutionary groupings.
- Gross and histological anatomical similarities within, and differences between, francolin and spurfowl syrinxes support their division into 'true' francolins and spurfowls.
- Francolins are on a 'perch' of the gamebird evolutionary tree, distant from spurfowls, and are more closely related to the Asiatic francolins, the Junglefowl *Gallus gallus* and the Bamboo Partridge *Bambusicola thoracica*.
- Spurfowls, on the other hand, group with quails and a range of 'partridges' (for example, the Chukar Partridge *Alectoris chukar*) from the Northern Hemisphere.

- Since African francolins link with Asiatic francolins, including the Black Francolin *Francolinus francolinus*, the type species of the genus (the first francolin described to science), they have retained the common name 'francolin'.
- The five 'Old World' African partridges are members of the family Phasianidae and of four genera:
  - *Alectoris* Kaup, 1829;
  - *Ammoperdix* Gould, 1851;
  - *Xenoperdix* Dinesen, Lehmberg, Svendsen, Hansen and Fjeldså, 1994; and
  - *Margaroperdix* Reichenbach, 1853.
- Although the Stone Partridge *Ptilopachus petrosus* and Nahan's Partridge *P. nahani* (previously Francolin/Spurfowl) are most closely related to the New World quails (Odontophoridae), they are placed in this section to be with the other partridges.

Conservation

Francolins are generally less opportunistic in their use of altered and transformed habitats than spurfowls and are wilder, preferring natural habitats over the extensive use of agricultural lands. They are also less abundant overall and are only in some cases available at levels that can sustain wing-shooting, such as the Grey-winged Francolin *Scleroptila afra*. The general threats which impact on their populations are the injudicious use of fire, overgrazing by livestock and expanding rural settlements. Four of the 20 species covered in this section are threatened: Udzungwa Forest Partridge *Xenoperdix udzungwensis*, Rubeho Forest Partridge *X. obscuratus* and Nahan's Partridge are listed as Endangered and Ring-necked Francolin *S. streptophora* is listed as Near Threatened (The IUCN Red List of Threatened Species v 2015-4), although this last listing may still be conservative.

Crested Francolin
Francolin huppé

Dendroperdix sephaena

Perdix sephaena Smith, A, 1836, Marico River, Limpopo Province, South Africa

Crested Francolin subspecies *D. s. sephaena* from southern Africa (photo by Ian White)

Crested Francolin subspecies *D. s. grantii* from Tarangire National Park, Tanzania (photo by Sergey Dereliev)

Crested Francolin subspecies *D. s. rovuma* from Mphingwe Camp, Catapu, Mozambique (photo by Maans Booysen)

Classification

Crested Francolin are related to Grey Francolin *Francolinus pondicerianus* and Swamp Francolin *F. gularis* of Asia. Recent research suggests that all three should be reassigned to the genus *Ortygornis*. Five subspecies are recognised:

D. s. grantii (Hartlaub, 1866) in southern Sudan and western Ethiopia to north-central Tanzania

D. s. spilogaster (Salvadori, 1888) in eastern Ethiopia, Somalia and north-eastern Kenya

D. s. zambesiae (Mackworth-Praed, 1920) in west-central Mozambique to north-western Namibia and southern Angola

D. s. sephaena (Smith, A, 1836) in south-eastern Botswana, eastern Zimbabwe, southern Mozambique and north-eastern South Africa

D. s. rovuma (Gray, GR, 1867) from Kenyan to central Mozambique, is streaked onto the belly and flanks, commonly known as Kirk's Francolin

Suggestions are that *D. s. grantii* be afforded full species status, including *D. s. spilogaster*, and similarly that *D. s. rovuma* be regarded as a full species.

Description

- They have a bantam-like build and characteristic white striped-and-streaked head, extending down the centre of the back and breast feathers.
- Their neck, upper breast and back are reddish-brown, contrasting with the paler belly.
- The rump is yellowish-brown and the vent pinkish, while the tail feathers are black, and are conspicuous when the tail is fanned in flight.
- They have a black bill and purplish-red legs and feet.
- Males have a dark brown cap with a grey outer edge that contrasts with the broad white eyebrow stripe.
- Males have long, curved leg spurs.
- Females are similar, but less boldly marked.
- The back plumage is barred in females and speckled in males.
- Juveniles resemble adult females, but have paler upperparts and the white streaking on the feather shafts is broader and more clearly defined.
- Downy chicks have rufous-brown upperparts, with a broad dark streak on the crown and back, somewhat similar to spurfowl chicks.
- The side of the head has a dark blotched stripe through the eye.
- The underparts are buffy-white, darker on the breast.

Distribution and habitat

- They are distributed from south-eastern Sudan, central Ethiopia and north-western Somalia, south through Kenya, Uganda, Tanzania, Mozambique, southern Zambia, Zimbabwe to southern Angola, northern Botswana and Namibia to north-eastern South Africa.
- Their preferred habitats are thickets and woodlands, particularly thornveld, often with sparse ground cover, as well as bush-encroached savanna.

Habits

They are territorial and secretive, coveys of 2–5 birds being seen along roadsides and in openings in woodlands. During the heat of the day, they retire to dense undergrowth, where the sound of their scratching in leaf litter discloses their presence. When disturbed, they prefer to run rather than to flush, scampering into dense vegetation. When flushed, they fly low and fast, weaving between trees. Unlike other grassland francolins, they will perch in trees to avoid ground predators. They also often roost on a horizontal branch within the crown of a tree. The advertisement call, uttered mostly in the early morning and less so at sunset, is a closely synchronised, repeated '*kee, kik, kerrik*' antiphonal duet.

Food and feeding

They feed on corms and bulbs, green shoots, leaves, fruits and berries of *Grewia occidentalis*, *Rhus lancea*, mistletoe seeds (Loranthaceae) and *Ziziphus mucronates*, and insects during the breeding season. They are also known to feed on tree resin (*Senegalia mellifera* and *S. caffra*), which they glean from the bark while perching in trees. They will ingest seeds and invertebrates from the droppings of large herbivores, such as African elephant *Loxodonta africana*, buffalo *Syncerus caffer* and rhinoceros (Rhinocerotidae).

Breeding

They are monogamous, breeding during October to March in the south and during April to June in East Africa. The nest is a scrape in the ground under dense, low bush, lined with grass and leaves. The clutch of 4–9 rounded eggs are unmarked cream or pinkish-buff and measure 39.7 x 30.4 mm. Incubation by the female lasts 20–22 days. Males guard the nest area, warning the incubating female of impending danger by uttering a soft '*koo woo koo woo kirr kirr kirr kirr*' call.

Status and conservation

Although their densities can vary between sites, they are not threatened, being neither rare nor vulnerable. The only potential threat is the clearing of bush for domestic fuel, agriculture and grazing.

Coqui Francolin
Francolin coqui

Peliperdix coqui

Perdix coqui (Smith, A, 1836), near Kurrichane, Zeerust, North West Province, South Africa

Male Coqui Francolin subspecies *P. c. hubbardi* from Kenya (photo by Jacques Pitteloud)

Calling male Coqui Francolin subspecies *P. c. coqui* from southern Africa (photo by Hugh Chittenden)

Female Coqui Francolin subspecies *P. c. coqui* (photo by Maans Booysen)

Classification

They are most closely related to Schlegel's Francolin *P. schegelii* and White-throated Francolin *P. albogularis*. Four subspecies are recognised:

P. c. spinetorum (Bates, 1928) in Mauritania, and Mali to northern Nigeria

P. c. maharao (Sclater, WL, 1927) in southern Ethiopia, central and eastern Kenya

P. c. hubbardi (Ogilvie-Grant, 1895) in western and southern Kenya, and northern Tanzania

P. c. coqui (Smith, A, 1836) in Gabon, the Democratic Republic of the Congo to southern Kenya and Uganda, south to South Africa

Description

- Coqui Francolin are sexually dimorphic. Males have a plain mustard-yellow head with a darker chestnut crown, and a black-and-white barred breast and shoulders.
- The orange-buff tail of both sexes is conspicuous in flight.
- The legs and feet are rich yellow.
- Males have sharp leg spurs.
- Females resemble Shelley's Francolin *S. shelleyi*, but have a broad white stripe above the eye and onto the neck.
- Juveniles resemble adult females, but are paler above, with rufous-buff mottling.
- Their underparts are buffy, with faint black barring.
- Downy chicks are rufous-brown above, with a broad dark streak on the crown and back, somewhat similar to spurfowl chicks.
- The sides of the head have a blotched dark streak through the eye.
- The underparts are buffy-white.

Distribution and habitat

- They are widespread in Africa, but patchy in West Africa with isolated populations.
- They inhabit tall grasslands (50–100 cm tall), savanna or well-grassed miombo (*Brachystegia*) and less so mopane (*Colophospermum mopane*) woodlands, and in drier country on sand dunes with good bush cover.
- They do not regularly venture far into agricultural lands.

Habits

They occur in pairs or small coveys of up to 12 individuals. Coveys are territorial and are difficult to see, reluctant to flush, choosing to run low through grass cover to avoid detection. In bushy habitat, they will flush from the far side of a bush. Once flushed, they fly relatively far, dodging between trees, and seldom flush a second time. When approached by a vehicle, they crouch, relying on their cryptic camouflage, resulting in many road-kills. They have two characteristic calls: a repetitive, two-syllable '*co-qui*' or a loud, tinny trumpet-like '*ter, ink, ink, terra, terra, terra, terra, terra*', with the second or third notes being loudest and the last notes fading away. Calling birds often perch on a termite mound, tree stump or boulder. In north-west Africa, the tempo of this call becomes truncated.

Food and feeding

Their diet consists of fewer underground corms and bulbs than other grassland francolins, with more seeds and small fruits consumed. They forage by using a sideways motion of the bill. Their breeding season diet includes beetles, grasshoppers, mantises and winged ants. They are known to drink dew or raindrops from grass blades and are probably not dependent on standing water.

Breeding

Breeding can occur during most months of the year, but mostly during the rainy season. The nest is a shallow hollow in dense cover lined with grass and leaves. The clutch of 4–5 oval, plain, cream or pinkish-buff eggs measure 32.5 x 27.3 mm. Incubation is by the hen, but the incubation period is not known. Chicks can fly at 7–10 days.

Status and conservation

There is no evidence that the distribution of this species has changed significantly in recent times. Potential threats are changes in grass cover as a result of frequent fires and continuous grazing by livestock, cultivation for crops and motor vehicle road-kills.

White-throated Francolin
Francolin à gorge blanche
Peliperdix albogularis
Francolinus albogularis (Hartlaub, 1854), The Gambia

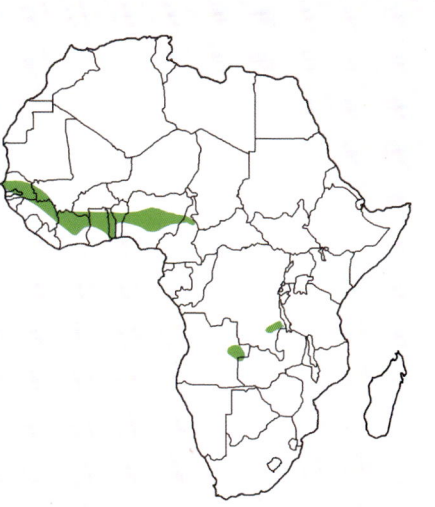

White-throated Francolin subspecies *P. a. buckleyi* from the Benoue National Park, Adamawa Province, Cameroon (photo by Nik Borrow)

White-throated Francolin subspecies *P. a. albogularis* from the Mole National Park, Ghana (photo by Adam Riley)

White-throated Francolin subspecies *P. a. albogularis* from Nigeria (photo by Tasso Leventis)

Classification

They are most closely related to Coqui Francolin *P. coqui* and Schlegel's Francolin *P. schegelii*. There are three subspecies recognised:

P. a. albogularis (Hartlaub, 1854) in Senegal and The Gambia to Ivory Coast

P. a. buckleyi (Ogilvie-Grant, 1892) from eastern Ivory Coast to northern Cameroon

P. a. dewittei (Chapin, 1937) in south-eastern Democratic Republic of the Congo, north-western Zambia and eastern Angola

Description

- The White-throated Francolin is similar to the Coqui Francolin *P. coqui*, but has a distinct white throat and no black barring on the hind neck and underparts of the male.
- They are best identified by their small size and buffy head, with grey crown and ear coverts, white supercilium, and plain buff unbarred underparts.
- The male of the distinctly different southern race *P. a. dewittei* has a chestnut breast band.
- Females are duller, with fine dark bars on their breast and flanks.
- The chestnut wings and tail of both sexes are conspicuous in flight.
- Juveniles have more barred underparts.
- Downy chicks are buff brown above and paler below.

Distribution and habitat

- They are distributed across West Africa from Senegambia, Guinea and south-western Mali to central Nigeria and northern Cameroon.
- The status of separate isolated populations in south-eastern Democratic Republic of the Congo and eastern Angola are uncertain.
- They prefer grassy plains and open savannas, often in recently burnt areas, and also venture into disturbed areas and the edges of cultivated lands.

Habits

They occur in pairs or small coveys. They are shy and sit tight or reluctantly scurry off through the grass sward if disturbed. If flushed, they fly fast and low with their neck extended. Their advertisement call is a far carrying, high-pitched trumpet-like '*ter-ink-inkity-ink*', somewhat similar to Coqui Francolin, but faster and slightly higher pitched, with up to eight notes, rather than four to six.

Food and feeding

Their diet consists mostly of grass seeds and green plant material as well as grasshoppers, termites, beetles, ants and other insects.

Breeding

They breed during September and October in Senegambia and during June in Nigeria. Their nest is a scrape lined with grass and leaves. The clutch of 4–7, usually 6, rounded buff to pale brown, slightly glossy and finely pitted brown-speckled eggs measure 32.4 x 26.4 mm.

Status and conservation

They are uncommon to rare across most of their range, but are widespread and relatively tolerant of disturbed habitats.

Schlegel's Francolin
Francolin de Schlegel

Peliperdix schlegelii

Francolinus schlegelii (Heuglin, 1863), Bongo River (Bussere River), Bahr el Ghazal, Sudan

LEFT Female Schlegel's Francolin with black-blotched breast bars (photo by Nik Borrow)

BELOW Male Schlegel's Francolin with finely barred breast (photo by Nik Borrow)

Classification

They are most closely related to Coqui Francolin *P. coqui* and White-throated Francolin *P. albogularis*. They are monotypic with no subspecies recognised.

Description

- Schelgel's Francolin is the Central African counterpart of the White-throated Francolin, but both sexes without a white throat and with their rusty-orange or rufous-yellow head look more like the male Coqui Francolin.
- However, they differ from Coqui Francolin by having a grey-brown crown and eye-stripe line from the lores to the ear coverts.
- The underparts are distinctly barred black on buff-white.
- Their bill is black with a yellow base and their legs are yellow.
- Females have browner upperparts which are less richly coloured than males and the belly is less heavily barred.
- Juveniles are similar to adult females, but have rufous-buff barring on the mantle and scapulars.
- The downy chick is undescribed.

Distribution and habitat

- They occur from west-central Cameroon (Adamawa Plateau), through northern Central African Republic and southern Chad to western South Sudan (Bahr el Ghazal).
- They are localised and uncommon, inhabiting rank grasslands in well-wooded savanna, particularly where the tree *Isoberinia doka* is common.
- They avoid human settlements, but will forage on the edges of cultivated lands.

Habits

They occur in pairs or small coveys. When flushed, their flight it slow, short and silent. They roost on the ground, huddled together with their heads facing outwards. Their repeated trumpet-like '*ter, ink, terrra*' call is similar to that of the Coqui and White-throated Francolins, but is faster and slightly lower pitched.

Food and feeding

Their diet consists of grass seeds, leaves, agricultural grain and caterpillars.

Breeding

They are monogamous and breed from September to November, at least in South Sudan. Their nest is a shallow hollow in the ground lined with grass and leaves. The clutch of 2–5 smooth cream-coloured eggs measure 35.5 x 26.0 mm. Their incubation period is unknown.

Status and conservation

They are generally localised and uncommon to rare, and their status and threats are unknown.

Shelley's Francolin
Francolin de Shelley
Scleroptila shelleyi

Francolinus shelleyi Ogilvie-Grant, 1890, Hartley Hills, Umfuli River, Mashonaland, Zimbabwe

ABOVE Calling Shelley's Francolin subspecies *S. s. shelleyi* (photo by Andre Botha)

TOP RIGHT Shelley's Francolin subspecies *S. s. uluensis* from Kenya (photo by Jacques Pitteloud)

RIGHT Ruffled Shelley's Francolin (photo by Andre Botha)

Classification

Shelley's Francolin are most closely related to Orange River Francolin *S. levalliantoides*, Archer's Francolin *S. gutturalis* and Moorland Francolin *S. psilolaema*.

Four subspecies are recognised:

S. s. uluensis (Ogilvie-Grant, 1892) in central and southern Kenya and northern Tanzania

S. s. macarthuri (van Someren, 1938) in the Chyulu Hills, south-eastern Kenya

S. s. shelleyi (Ogilvie-Grant, 1890) in southern Uganda, south through Tanzania, Zambia, southern Malawi, Zimbabwe and Mozambique to north-eastern South Africa and Swaziland

S. s. whytei (Neumann, 1908) in south-eastern Democratic Republic of the Congo, northern Zambia and northern Malawi

Description

- They are difficult to distinguish from Orange River Francolin, Archer's Francolin and Red-winged Francolin *S. levaillantii*, as they have the same white throat.
- However, the black necklace is thinner than that of Red-winged Francolin and broader than that of Orange River Francolin and Archer's Francolin.
- The white lower breast feathers are boldly marked with two or more black bands (these feathers are buff to rufous in the other three species).
- The legs and feet are yellow.
- Males have leg spurs.
- Juveniles resemble adults, but the black-and-white breast barring is irregular and the outer margins of the primary flight feathers are marked with buff.
- Downy chicks are rufous-brown above, with strong darker stripes on the crown and back, and buffy below.
- They have characteristic multiple, alternating black and buff stripes above and around the eye.

Distribution and habitat

- Their distribution is patchy from Uganda and Kenya through Tanzania, Zambia and Malawi to Mozambique and South Africa.
- They prefer open grasslands, wooded savanna, thornveld and open areas in miombo (*Brachystegia*), mopane (*Colophospermum mopane*) and other woodlands, often on stony terrain or among rocky outcrops.

- They prefer taller sweetveld in the east, whereas Red-winged Francolin inhabit shorter sourveld dominated by red grass *Themeda triandra*.
- Both species are separated from Orange River Francolin, which occur in drier, tall, sparse grasslands of western southern Africa.

Habits

Usually in sedentary pairs or small coveys of 6–8 birds, they sit tight when approached, but will run through grasslands to avoid being flushed. Where the home range includes moist sites, they will often use marshes and sponges to roost. They generally fly to their overnight roosting site, thus avoiding being tracked by predators. They often use the edges of cultivated lands, but do not spend much time within agricultural lands. The musical call consists of two low notes followed by two higher notes, '*I'll drink YER BEER*', repeated three or four times, usually from a vantage point such as a termite mound or rock. This call is not as high pitched and is more deliberate than that of Orange River Francolin, being uttered at a much slower pace. They also have a shrill flush call.

Food and feeding

They forage by digging for corms and bulbs of plants with their bill. Their diet also includes seeds and grains, and is supplemented with invertebrates during the breeding season.

Breeding

They are monogamous, solitary nesters. In southern Africa, peak nesting occurs from October to January, while in East Africa it is during the dry season (March–July). The nest is a relatively deep scrape in dense grassland or under a bush, well lined with grass and roots. The clutch of 4–5 (occasionally 6 or 7) sometimes pitted oval eggs are buff or whitish and measure 38.0 x 31.1 mm. Incubation by the female lasts for 20–22 days. Chicks can flutter-fly at 12 days and can fly strongly at 5 weeks.

Status and conservation

Although they are not threatened, there is some evidence of a reduction in their range due to grassland deterioration resulting from continuous grazing by domestic livestock and frequent burning. Fires in these grasslands destroy both nests and chicks, and expose nesting sites to severe hailstorms.

Grey-winged Francolin
Francolin à ailes grises
Scleroptila afra

Francolinus africanus Latham, 1790, Cape Province, South Africa

ABOVE Female Grey-winged Francolin (photo by Jessie Walton)

RIGHT Calling male Grey-winged Francolin (photo by Andre Botha)

FRANCOLINS & PARTRIDGES

ABOVE Grey-winged Francolin nest and clutch (photo by Rob Little)

Incubating female Grey-winged Francolin (photo by Andre Botha)

One day old Grey-winged Francolin chick showing typical fine head-stripes of francolin chicks (photos by Rob Little)

Classification

Grey-winged Francolin have previously been considered conspecific with Moorland Francolin *S. psilolaema* of East Africa. But, they are geographically separated and differ markedly in appearance, thus rather being considered endemic to their respective ranges.

They are monotypic. A suggested subspecies *S. a. proximus* (Clancey, 1957) of the Drakensberg region is not supported.

Description

- They are a medium-sized francolin, and the sexes are not distinguishable other than that the adult male has short, robust leg spurs.
- They differ from the other francolins in the red-winged group by having very little red on their wings, and they have a distinctive grey-freckled throat, not white or buff.
- Juveniles resemble adults, but have duller plumage and white on the throat.
- Downy chicks are rufous-brown above, with darker streaking on the crown and back.
- The sides of the head, above and below the eye, have multiple fine black-and-white stripes.
- The underparts are buffy-white, washed with rufous on the breast.

Distribution and habitat

- They are endemic to South Africa and Lesotho.
- They thrive in the highland grasslands of the Drakensberg, from Mpumalanga along the KwaZulu-Natal escarpment and the Lesotho Maluti range to the Eastern Cape.
- They prefer patches of sparse, short grass on shallow, igneous soils on ridge tops and shelves, where grasslands are lightly grazed by wild herbivores and burnt less frequently than every two years.
- Their distribution is patchier and their numbers are lower in the Western Cape, where they occur in Karoo scrub and coastal renosterbos down to sea level.
- They are more tolerant of heavy grazing than Red-winged Francolin *S. levailantii*.

Habits

They are sedentary and territorial, with non-breeding coveys consisting of 4–20 individuals. Coveys usually sit tight if disturbed, but will run if the cover is too sparse. They seldom fly except when flushed, or when flying to their overnight roost site in the late afternoon. The covey roosts on open ground, huddling together. They have a high-pitched, whistling advertisement call '*pip-pip-pip-pip pi-pip wi-pleeu*', which is most often heard at sunrise, and also a flush call *'pre-pre-pre-pre-preeu-preeu-preeu-preeu'* when a covey is disturbed. The *'wi-pleeu'* phrase of the advertisement call is repeated in series and is often punctuated antiphonally by another bird giving a *'pipeeu'* call.

Food and feeding

Their diet consists of underground corms, rhizomes and bulbs of the sedge and iris families (*Cyperus*, *Mariscus* and *Moraea* species) during the non-breeding season (February–September), and of invertebrates during the chick-rearing period (October–January). Seeds are also an important part of their autumn diet.

Breeding

Although breeding peaks during August and September, the breeding season is prolonged (from August–March) in the summer-rainfall grasslands and contracted (July–December) in the winter-rainfall region of the Western Cape. The nest is a scrape under a grass tuft, lined with grass and occasionally feathers. The average clutch of five to six yellowish-brown eggs, sometimes lightly speckled with brown and slate, measure 39.9 x 30.1 mm. Incubation by the female is for 21–22 days.

Status and conservation

Nest predation by Yellow Mongoose *Cynictis penicillata*, crows *Corvus* species and Common Egg-eater *Dasypeltis scabra* snakes is a concern, particularly in heavily grazed grasslands. Other causes of nest failure are fires and trampling by livestock. Although a fine-scale fire mosaic favours grassland francolins, burning should cease before the end of August. A sustainable Grey-winged Francolin wing-shooting industry is a viable commercial agricultural by-product in the Eastern Cape, South Africa, and provides an incentive for conservation of these grasslands.

Orange River Francolin
Francolin de L'Orange
Scleroptila levalliantoides

Perdix levaillantoides (Smith, A, 1836), upper reaches of Orange River, South Africa

ABOVE Female Orange River Francolin subspecies *S. l. jugularis* (photo by Cliff Dorse)

ABOVE Male Orange River Francolin subspecies *S. l. levalliantoides* (photo by Andre Botha)

LEFT Male Orange River Francolin subspecies *S. l. pallidior* (photo by Ian White)

Classification

Their closest relative is Archer's Francolin *S. gutturalis* of East Africa. Three subspecies are recognised:

S. l. levalliantoides (Smith, A, 1836) in southern Botswana, north-western and central South Africa

S. l. jugularis (Bötikofer, 1889) in south-western Angola

S. l. pallidior (Neumann, 1908) south of the Cunene River, in north-central Namibia and western Botswana

Recent research indicates that *S. l. jugularis* may be recognised as a separate monotypic species in the future.

Description

- They have a white throat and a characteristic black necklace around it, with a narrow black-and-white band behind the eye, which runs down either side of the neck, broadening, but does not meet on the throat.
- The upper breast and flanks are buff with red-brown blotches, while the belly is plain buff.
- The flight feathers are distinctly red in flight.
- The legs are yellow and males have leg spurs.
- Juveniles are similar to adult females, but the black-and-white banding on the head and neck is less well defined and the underparts are barred irregularly with black and buff.
- Downy chicks have reddish-brown underparts, with dark brown streaks on the crown and mid-back, with the ones on the back being flanked by buffy streaks.
- The sides of the head are buffy, and the brown eye has several black-and-white stripes above and below it.

Distribution and habitat

- They have a patchy distribution, with separate populations occurring in north-eastern and central Namibia, extending marginally into south-western Angola.
- From Namibia, their range extends through much of south-central Botswana into west-central South Africa.
- They inhabit mostly flat landscapes, often on sandy soils, except in Namibia, where they favour hilly terrain.
- Typical habitats are open, sandy grasslands of the dry savannas and the central Kalahari in the east and wooded, bushy grasslands in the west.

- Although typically a grassland species, they favour patches of moderately dense cover and often frequent the edges of cultivated lands.

Habits

They are usually seen in pairs or coveys of 3–5 birds. When approached, they sit tight and then run into thicker cover. Flushed birds will gain height before peeling off over trees. They dust-bath and roost in flat, open patches. Their repeated advertisement call '*kibitele, kibitele*' is similar to that of Shelley's Francolin *S. shelleyi*, but is more rapid and higher pitched, with the accent on the third note. Calling birds often perch on a conspicuous object such as a rock. Most calling occurs in the early morning and evening, particularly during the early breeding season. Other subdued, grating calls are used to communicate between individuals in a covey at close quarters or to indicate alarm at the approach of a predator or other threat.

Food and feeding

Their diet comprises corms, bulbs, seeds and some green shoots, supplemented with leaf bugs, beetles, ants, termites and grasshoppers during the breeding season.

Breeding

They are monogamous solitary nesters. Although breeding can be at any time of the year, peak breeding is during June and July in northern Namibia and northern Botswana, March and April in central Namibia and south-eastern Botswana, and August to October in South Africa. The nest is a scrape, usually in dense grass cover and lined with dry grass. The clutch of 5–8 pale coffee-coloured or yellowish oval eggs, sometimes lightly speckled with brown, measure 36.5 x 28.7 mm. Incubation by the female lasts for 20–21 days. Chicks can flutter-fly at 12–14 days and accomplished flight is at 5–6 weeks.

Status and conservation

Like some other grassland francolins, they are susceptible to habitat manipulation, particularly where poorly managed grazing has resulted in degraded grasslands. A few consecutive years of overgrazing, even on otherwise well-managed nature reserves, can cause local extinctions.

Archer's Francolin
Francolin d'Archer
Scleroptila gutturalis

Scleroptila gutturalis (Rüppell, 1835), north-eastern Africa

LEFT Archer's Francolin (photo by János Oláh)
BELOW Archer's Francolin (photo by Callan Cohen)

Classification

This species' closest relative is the Orange River Francolin *S. levalliantoides* of southern Africa, from which it is now regarded as a separate full species.

The two subspecies of Archer's Francolin, *S. g. lorti* (Sharpe, 1897) and *S. g. archeri* (Sclater, WL, 1927), of Ethiopia, South Sudan, Somalia and northern Uganda, are likely to be combined under *S. gutturalis* to form one monotypic species in the future.

Description

- The Archer's Francolin was previously considered a race of the Orange River Francolin.
- They are darker than the Moorland Francolin *S. psilolaema*, lacking streaking on their white throat, and they have paler, less rufous underparts.
- Their buff, not white, line behind the eye and paler underparts, which are lightly streaked black, also distinguish them from the Shelley's Francolin *S. shelleyi*.
- Juveniles are similar to adult females, but are duller with less defined facial striping and gorget pattern.
- Downy chicks are not described.

Distribution and habitat

- They are a locally common endemic to north-eastern Africa, in Ethiopia, southern South Sudan, northern Somalia and northern Uganda.
- Their preferred habitats are open and slightly wooded grasslands and boulder-strewn scrub-covered hill slopes.

Habits

They usually occur in pairs or small coveys and are shy and elusive. Their presence is usually only detected by their call, mostly in the early morning and late afternoon. Their repeated '*ki-bi-til-ee*' call is similar to that of Shelley's Francolin, but is faster, shorter and more abrupt. They also have a high-pitched, squealing flight call when flushed.

Food and feeding

Their diet consists mostly of bulbs and corms (*Moraea* spp.), seeds, berries and fallen grain, as well as bugs, termites, beetles and grasshoppers.

Breeding

They are monogamous, nesting in a well-concealed scrape under a grass tuft. Laying dates are recorded as February, March and August for Ethiopia. The clutch of 5–8 pale pink to yellowish-brown eggs, sometimes with brown speckling, measure 41.2 x 31.2 mm.

Status and conservation

They are generally not threatened, but although they are locally common their status is uncertain. Heavy grazing by domestic livestock and annual burning of the grasslands probably impact on their habitat quality.

Red-winged Francolin
Francolin de Levaillant

Scleroptila levaillantii

Perdix levaillantii (Valenciennes, 1825), Swellendam, Western Cape Province, South Africa

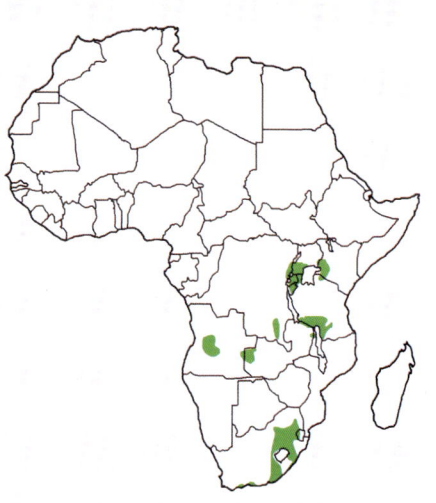

TOP RIGHT Male Red-winged Francolin (photo by Maans Booysen)

TOP LEFT Red-winged Francolin with a flying termite (photo by Andre Botha)

LEFT Incubating female Red-winged Francolin (photo by Ian Little)

FRANCOLINS & PARTRIDGES

Classification

Three subspecies are recognised:

S. l. kikuyuensis (Ogilvie-Grant, 1897) in Angola, western Zambia, eastern Democratic Republic of the Congo, eastwards to western and central Kenya

S. l. crawshayi (Ogilvie-Grant, 1896) in northern Malawi

S. l. levaillantii (Valenciennes, 1825) in south-western Malawi, north-eastern Zambia and eastern South Africa

Description

- They are the largest of the red-winged group, with prominent rufous markings on the side and hind neck and a broad black-and-white collar, but lack a black border to the white throat.
- They have a large, robust bill, which allows for digging out relatively large bulbs.
- Males have relatively small leg spurs.
- Juveniles resemble adults, but are duller, and the black-and-white barring on the lower neck and upper breast is less distinct.
- Downy chicks are rufous-brown above, with dark brown streaks on the crown and mid-back, the latter flanked by buffy streaks.
- The sides of their head are buffy, with multiple black-and-white stripes above and below the eye.
- The underparts are buffy, but somewhat darker on the breast.

Distribution and habitat

- Southern African populations are isolated from those in Central and East Africa.
- In southern Africa, they occur in the southern Mpumalanga highlands, the highlands and Midlands of KwaZulu-Natal, the eastern Free State and the Eastern Cape, with remnant populations along the southern coast of the Eastern and Western Cape.
- In the eastern parts of South Africa, they inhabit montane sour grasslands, dominated by red grass *Themeda triandra*.
- They thrive in rank vigorous habitats, usually over deep soils.
- They are more sensitive to burning or grazing of grasslands than other members of the red-winged group.
- In heavily grazed and frequently burnt grasslands, they are restricted to wetland fringes and rocky outcrops.

Habits

They occur in relatively small coveys of 4–8 birds. They are rarely flushed, but are commonly heard calling from hill tops early in the morning and sometimes at dusk. When flushed, they explode out of the grass with a flurry of wings. The escape flight usually ends in a long glide before dropping into dense grass or sedge cover, seldom to flush again. The distinctive, high-pitched whistle call is rendered as '*chee-chee-chee-chee kla-CHEE-choo*', accelerating into repeated '*klik-a-chee-choo, klik-a-chee-choo*' phrases.

Food and feeding

They feed primarily on bulbs or corms, scraping with the heavy bill around rocky outcrops and wetland fringes where favoured food plants, such as *Hypoxis*, *Rhodohypoxis*, *Hesperantha*, *Oxalis*, *Gladiolus* and *Moraea* species, abound. Their diet is supplemented with insects during the breeding season.

Breeding

Peak breeding largely follows the early rains. The nest site is often in close proximity to wetland edges within two- to three-year-old grass. The nest is a shallow scrape between grass tufts lined with live and dead grass. The clutch of 5–8 creamy yellow eggs, usually faintly spotted with dull brown, measure 40.1 x 32.3 mm. Incubation by the female lasts 22 days. Chicks are precocial and are cared for by both parents.

Status and conservation

There are numerous examples of local declines in their numbers and areas of confirmed absence in apparently suitable habitat in southern Africa. They are sensitive to annual burning and overgrazing, as well as to unburned moribund grassland. Biennial burning and light grazing by wild herbivores provides important sanctuaries for this species. Other major threats are expanding commercial forestry and the damming and draining of wetlands.

Finsch's Francolin
Francolin de Finsch

Scleroptila finschi

Francolinus finschi Barboza du Bocage, 1881, Caconda, Benguella, Angola

Elusive Finsch's Francolin only represented here by camera trap images set by David Elizalde

Classification

Finsch's Francolin are most closely related to Shelley's Francolin *S. shelleyi*. They are monotypic with no subspecies recognised.

Description

- They can be considered as the localised counterpart of the red-winged francolin group, with duller, less patterned plumage.
- They are the only red-winged group member with no black-and-white stripe pattern on the face or neck, and they have a grey breast and belly.
- The centre of their belly is buff with chestnut blotching and it is essentially unbarred.
- The sides of the face and border of the throat are ochre as in Red-winged Francolin *S. levaillantii* but this is less extensive on the hind neck.
- This species is also long-billed like Red-winged Francolin.
- They are distinguished from Coqui Francolin *Peliperdix coqui* by their larger size and the male has less orange on the head.
- They differ from the other grassland francolins by their yellowish legs and black bill.
- The primaries and outer secondaries are only slightly rufous in flight.
- Both the juveniles and downy chicks are undescribed.

Distribution and habitat

- The most extensive population is in west-central Angola, with localised, patchy populations in south-eastern Gabon, Congo and south-western Democratic Republic of the Congo.
- They inhabit mainly short-grass savannas, *Brachystegia* woodland, forest edges and bare slopes above the timberline in mountainous terrain.
- They have also been recorded feeding in recently burnt areas.

Habits

They are usually seen in pairs. Their call is a loud '*wit-u-wit*' duet, usually heard at dusk, somewhat similar to that of Shelley's Francolin.

Food and feeding

Their diet includes seeds, beetles and insect larvae.

Breeding

Probably monogamous, they breed during January, March and July in the DRC and during June and July in Angola. The nest is placed within vegetation on the ground. Their clutch consists of, on average, 5 eggs, which are plain light brown. Their incubation biology is unknown.

Status and conservation

Although the bulk of their range is in Angola, they are sparsely distributed and uncommon to rare. There is little known about their potential threats, but similar to most grassland francolins, habitat degradation plays a role.

Moorland Francolin
Francolin montagnard
Scleroptila psilolaema

Francolinus psilolaemus Gray, GR, 1867, Shoa, Ethiopia

Moorland Francolin subspecies *S. p. elgonensis* with plain off-white throat, white-spotted black throat-band and rufous breast with full black spots. Possible future full species Elgon Francolin *S. elgonensis*. (Photo on Mount Kenya by Callan Cohen & Michael Mills www.birdingafrica.com)

Moorland Francolin subspecies *S. p. psilolaema* (photo by Nigel Redman)

Moorland Francolin subspecies *S. p. psilolaema* (photo by Otto Schmidt)

Classification

They are closely related to Grey-winged Francolin *S. afra*. There are two subspecies recognised:

S. p. psilolaema (Gray, GR, 1867) in central and southern Ethiopia

S. p. elgonensis (Ogilvie-Grant, 1891) in eastern Uganda to central Kenya

Description

- The Moorland Francolin is generally darker than the Archer's Francolin *S. gutturalis*, with a buff variably spotted or streaked, not white, throat and more rufous underparts with small black spots across the upper breast.
- They have a rather plain, buffy belly, not barred like that of the Shelley's Francolin *S. shelleyi*.
- The wings show fairly extensive rufous in flight, approaching that of the Red-winged Francolin *S. levaillantii*.
- They are the only member of the red-winged francolin group with black bars at the tips of their primaries.
- The Kenyan sub-species *S. p. elongensis* have richer rufous underparts.
- Juveniles and downy chicks are undescribed.

Distribution and habitat

- They are endemic to north-eastern Africa, occurring in the montane areas of central and south-eastern Ethiopia, far western Uganda and western Kenya.
- They are localised and uncommon on Mount Kenya, the Aberdares and Mau Narok in Kenya, but more common on Mount Elgon in both Kenya and Uganda.
- Their preferred habitats are montane heath moorland and grasslands, adjacent pastures and moist grassy areas sometimes near ponds, and are rarely found below 2 400 m.

Habits

They are usually found in pairs or small coveys which are shy. Calling coveys will repeatedly answer one another with the typical advertisement call. Their advertisement call is a repeated short and unhurried '*trich-chi-che'e*', very similar to that of Shelley's Francolin. They also utter a high-pitched squealing when flushed.

Food and feeding

Similar to other grassland francolins of the red-winged group, they feed mostly on corms, bulbs and roots. They forage by excavating with their bill in damp grassy areas and amongst sedges, often around montane pools and sponges.

Breeding

They are monogamous and breed mostly in the first half of the year. The nest is a scrape in open grassland or heath. The clutch is usually 4–5 eggs. Their incubation biology is unknown.

Status and conservation

They are uncommon to rare in Kenya and locally more common in Ethiopia. However, they may be threatened by the reduction and degradation of high-altitude moorland and grassland habitats through overgrazing and agricultural cultivation. The heather moorlands have the highest density of livestock in Ethiopia, and habitat loss and disturbance, including trampling and damage to nesting sites, are probably a real concern.

Ring-necked Francolin
Francolin à collier

Scleroptila streptophora

Francolinus streptophorus (Ogilvie-Grant, 1891), Mangiki, Mount Elgon, Kenya

Ring-necked Francolin illustration by John Gerrard Keulemans – Catalogue of the birds in the British Museum. Volume 22 (no photograph available)

Classification

Ring-necked Francolin is an evolutionary basal member of the red-winged francolin group. They are a monotypic species with no geographic variation and thus no subspecies are recognised.

Description

- The Ring-necked Francolin is a distinctive member of the red-winged group of francolins.
- They are distinguished by their dark cap, broad white supercilium, chestnut face and neck, contrasting white throat and characteristic finely barred black-and-white breast collar.
- Females have a dark brown crown with a pale edge and rufous-brown barred upper parts, with broad pale shaft-streaks.
- They have a black bill and yellow legs.
- Juveniles and downy chicks are undescribed.

Distribution and habitat

- They have a patchy disjunct distribution in Uganda, Burundi, Rwanda, western Kenya and north-western Tanzania, with a further isolated population in the highlands of Cameroon.
- They inhabit open grasslands and lightly wooded areas on rocky hillsides.

Habits

They usually occur in pairs or small coveys, and keep themselves well hidden in sparse grass, running swiftly between and over rocks. They are shy and skulking, running for cover when disturbed, and are possibly one of the least photographed gamebirds in Africa. When flushed, they fly fast. They call mostly in the early morning and often from a prominent perch such as a termite mound. The call is a fluty dove-like cooing, followed by a piping trill. They also give a raucous alarm call in flight.

Food and feeding

Their diet consists of seeds and insects, but may also enter cultivated areas to feed on weeds, waste grains and insects.

Breeding

They are monogamous, breeding during the early rains (April) in Uganda and during the dry season (December–March) in western Kenya. The nest is a shallow scrape with little or no lining, placed close to a rock. The clutch of 4–5 greyish buff eggs have dark speckled pores. Their incubation biology is unknown.

Status and conservation

They are generally uncommon, but are fairly common in Tanzania. Rarer than previously believed, they have apparently declined in both range and abundance in Kenya, northwestern Uganda and Rwanda, and are now suspected to have a moderately small population. They have therefore been moved from a species of Least Concern to Near Threatened (The IUCN Red List of Threatened Species v 2015-4), but even this listing might still be conservative.

Forest Francolin
Francolin de Latham

Afrocolinus (Peliperdix) lathami

Francolinus lathami (Hartlaub, 1854), Sierra Leone

RIGHT Male Forest Francolin in leaf litter (photo by Andy Pugh)

BELOW LEFT Male Forest Francolin (photo by Tasso Leventis)

BELOW RIGHT Typical Forest Francolin nest on dry leaves between buttress roots of a forest tree (photo by Nik Borrow)

Classification

The Forest Francolin represents its own evolutionary trajectory from that of the other francolins and has thus been reassigned to the genus *Afrocolinus*. There are two subspecies recognised:

A. l. lathami (Hartlaub, 1854) in Sierra Leone to north-western Democratic Republic of the Congo and Angola

A. l. schubotzi (Reichenow, 1912) in western DRC to south-western South Sudan, western Uganda and north-western Tanzania

Description

- Also known as Latham's Forest Francolin, they are a small, dark forest species.
- They are best distinguished from Nahan's Spurfowl *Ptilopachus nahani* by their pale grey, not black, ear coverts and their yellow, not red, legs.
- They have a chestnut cap and a dark black throat.
- The upperparts are dark brown with fine, pale streaks, while the underparts are black with distinct white spots.
- Females are paler brown than males on the upperparts, with buffy cheeks and supercilium, and pale brown underparts with cream spots.
- Juvenile males have a mottled black-and-brown crown, with a white chin and throat, and the sides of the head and ear-coverts are brownish.
- Juvenile females have a broad, black-tipped crown and nape feathers.
- Downy chicks have dark chestnut-brown crown and back, a buff face with chestnut eye-stripe from behind the eye to the nape.

Distribution and habitat

- They are distributed in equatorial lowland forest from Sierra Leone to the Democratic Republic of the Congo, far northern Angola and western Uganda, with a small separated population in South Sudan.
- They inhabit the undergrowth of primary and occasionally secondary forest.

Habits

They are secretive and difficult to observe in their dense habitat, although they will venture into glades during wet weather. They are difficult to flush, sitting tight or running through the undergrowth. If flushed, they fly fast for a short distance. Their call, often heard at night, is a dove-like '*kwee, coo, coo* …' or '*krook-kroo-kroo-kroo* …' cooing, often repeated, and also various whistles and clucking contact calls.

Food and feeding

They forage by scratching through the leaf litter, sometimes even at night. Their diet consists mostly of invertebrates, including termites, ants, snails, beetles and other insects and their larvae. They also consume fruits, seeds and green leaves.

Breeding

Most of their breeding records are for December to April. The eggs are laid on open ground, among dry leaves between projecting forest tree root buttresses. The clutch of 2, rarely 3, elongate-ovate, thick-shelled dark buff or light rusty brown eggs measure 38.3 x 26.5 mm. Their incubation biology is unknown.

Status and conservation

They are baited with termites and grains to areas where they are trapped with snares and shot. They are also threatened by forest destruction.

Barbary Partridge
Perdrix gambra

Alectoris barbara

Perdix barbara (Bonnaterre, 1790), Morocco

Barbary Partridge (photos by Bill Baston above and Sergio Bianchi left)

Classification

Four subspecies of the Barbary Partridge are recognised:

A. b. koenigi (Reichenow, 1899) in north-western Morocco and introduced to the Canary Islands

A. b. spatzi (Reichenow, 1895) in southern Morocco to central and southern Algeria and southern Tunisia

A. b. barbara (Bonnaterre, 1790) in north-eastern Morocco to northern Tunisia

A. b. barbata (Reichenow, 1895) in north-eastern Libya and previously north-western Egypt

Description

- Similar to the other *Alectoris* partridges, their most distinctive features are their chestnut brown crown, grey throat, and broad chestnut-and-white speckled lower neck band, which expands backwards.
- They also have an extended grey supercilium, which runs down the back of the neck.
- Juveniles are pale yellow and lack the grey throat and neck band.
- Downy chicks are pale buff with a dark rufous crown and hind-neck stripe, and dark streaks above and through the eye.

Distribution and habitat

- They are a North African and Sardinian species extending from Morocco through northern Algeria, Tunisia and coastal Libya.
- There is a gap in their range in the central Libyan coast and a separate population occurs in the Ajjer Mountains in the south-eastern Algeria section of the Sahara Desert.
- They commonly inhabit rocky or stony hillsides and deserts up to 3 300 m, with dry grass, sparse scrub or open woodlands from coastal dunes up to the top of the Atlas Mountains during the boreal summer.
- They also inhabit cultivated lands and groves of palm, citrus, olives and *Eucalyptus* species.

Habits

They occur in pairs or small groups. When approached, they prefer to run rather than flush, particularly where they can run between vegetation. If flushed they fly fast, direct and far before landing. Their alarm call when flushed is a rapid '*ckukachew-chew-*

chew-chew' and the advertisement call is a harsh '*kakalik*', similar to that of the Chukar Partridge *A. chukar*. When calling, males stand erect on the tips of their toes. They roost on the ground, often under the shelter of rocks or vegetation. They are not dependent on fresh water.

Food and feeding

They forage in the early morning and evening, sometimes into the night. Their diet consists of seeds, the fruits of *Euphorbia* species, the succulent leaves of *Sosola, Lycium* and *Asparagus* species, which provide moisture. They also eat insects, particularly ants, and will take the waste grains in agricultural fields.

Breeding

They are monogamous and breed mostly from March to June. The nest is a shallow scrape, often in the shelter of a bush, sometimes lined with vegetation. The clutch of 6–20 eggs, usually 11, smooth, slightly glossy buff, with fine brown speckles measure 40.5 x 30.4 mm. Incubation is by the hen and lasts for 24–25 days, similar to other *Alectoris* species.

Status and conservation

Although they are common to abundant in remote or protected areas, there is evidence of localised declines due to hunting and habitat loss to agriculture. The population in north-western Egypt is almost certainly extinct because of hunting.

Sand Partridge
Perdrix de Hey

Ammoperdix heyi

Perdix heyi Temminck, 1825, Desert of Akaba, Arabia

ABOVE AND BELOW LEFT Male Sand Partridges (photos by Tasso Leventis)

BELOW Photo by Yael Shiff

Classification

Four subspecies are recognised:

A. h. heyi (Temminck, 1825) in Israel and Jordan south to the Sinai Peninsula, Egypt and western Saudi Arabia

A. h. nicolli Hartert, 1919 in northern Egypt east of the Nile River south to central Egypt

A. h. cholmleyi Ogilvie-Grant, 1897 in central Egypt also east of the Nile River south to north-eastern Sudan

A. h. intermedia Hartert, 1917 in southern Arabia and southern Oman

Description

- This is a small partridge weighing about 180 g.
- They have rufous outer tail feathers which are conspicuous in flight.
- Males are pinkish-buff with a dark head, chestnut throat and boldly barred rufous and black flanks.
- They have a distinct white patch behind the eye, and orange-yellow bill and legs.
- Females are sandy greyish-brown, with pinkish bars on the sides of the neck and subtle vermiculation.
- Juveniles are similar to adult females.
- Downy chicks are pale buff above and whitish below.

Distribution and habitat

- In Africa, they are distributed from north-eastern Egypt east of the Nile River south to north-eastern Sudan.
- Their preferred habitats are the slopes of broken rocky or stony hills with minimal scattered vegetation and lightly vegetated valleys with shrubs and sparse grass.
- They seldom venture into open stretches of dry, flat sandy desert.

Habits

They are usually in small groups of 2–5 individuals. When approached they prefer to squat motionless and are very cryptic. If disturbed, they run fast between vegetation and rocks. They rest in shade during the heat of the day. Although they are known to drink, they can go for extended periods without water. They are mostly silent, but at dawn and dusk will utter their loud '*kew-kew-kew*' call, and their wings make a rattling sound in flight.

Food and feeding

Their diet consists of berries, particularly of *Salvadora persica* and *Commiphora* species, grass seeds, corms, bulbs, green leaves and insects.

Breeding

They are monogamous and their laying season peaks in April. The nest is a shallow scrape typically placed next to a bush or rock and often lined with grass. Their clutch is usually 5–7 hard-shelled eggs which are pale buff and measure 37.0 x 26.7 mm.

Status and conservation

Little is known about their population status in Africa, but they are abundant in Arabia. Their occurrence in Eritrea and northern Ethiopia is uncertain.

Udzungwa Forest Partridge
Xénoperdrix de Tanzanie

Xenoperdix udzungwensis

Xenoperdix udzungwensis Dinesen, Lehmberg, Svendsen, Hansen and Fjeldså, 1994, Ndundulu Mountains in the Udzungwa Highlands in Tanzania

Male Udzungwa Partridges, immature above and adult right, from the Udzungwa Ndundulu Forest (photos by Louis André Hansen)

Classification

The Udzungwa Forest Partridge was only discovered in 1991, first noticed as a pair of strange feet in a cooking pot in a Tanzanian forest camp by scientists who were invited to dinner by local villagers (See Dinesen et al. 1994 in the Bibliography).

The Rubeho Forest Partridge *X. obscuratus* is treated as a separate species. They gave rise to the Eurasian and North American gamebirds. Their closest relatives are the Asian Hill Partridges of the genus *Arborophila*.

Description

- They are small, approximately 29 cm long (230 g), boldly barred, brownish partridges with a rufous face and throat, black forehead, olive-brown crown with orange-red supercilium, a narrow white necklace bordering the grey underparts blotched with black before the ochre vent.
- They have a diagnostic bright red bill, brown iris and yellow legs.
- Both sexes are similar, except in that the black forehead is most pronounced in the males.
- Juveniles vary according to age with a brown forehead and a dark bill which progresses to orange and finally to red.
- Downy chicks are undescribed.

Distribution and habitat

- They are endemic to central Tanzania where they are scarce in the Udzungwa Mountains.
- They are restricted to montane evergreen forests, mostly on ridges and steep slopes, with a thick understory, and in sub-montane forests with scattered *Cyperus* sedges and ferns.

Habits

They have been encountered in coveys of up to 13 individuals. They spend most of the day skulking in the forest understory, roost in trees at night and call mostly during the early morning. Their call is a whistled '*teedli teedli*' with high peeping notes.

Food and feeding

They forage among the leaf litter of the forest floor, scratching and turning over leaves with their bill. Their diet consists mainly of beetles, ants, flies, woodlice and seeds.

Breeding

The only information available is that breeding takes place following rains, from October to December.

Status and conservation

Because of their limited distribution they are vulnerable to habitat loss and snaring, facing a high risk of extinction in the wild. Their conservation status is therefore listed as Endangered (The IUCN Red List of Threatened Species v 2015-4).

Rubeho Forest Partridge
Xénoperdrix des Rubeho

Xenoperdix obscuratus

Xenoperdix obscuratus Fjeldså and Kiure, 2003, Mafwemiro Forest, northern Rubeho Mountains, central Tanzania

Illustration of Rubeho Forest Partridge female left, male right, by Jon Fjeldså

Only available images of Rubeho Forest Partridges captured by a camera trap placed by Francesco Rovero, MUSE – Science Museum, Italy

Classification

The Rubeho Forest Partridge was formerly considered to be a subspecies of the Udzungwa Forest Partridge *X. udzungwensis* of the Udzungwa Mountains. However, genetic analyses suggest a lack of recent gene flow between the two populations, which are separated geographically by only 100 km. They also differ in details of ornaments and colours of the face and tail, which are likely to function as visual signals. They are therefore now recognised to be sufficiently distinct to be treated as a separate monotypic species (See Bowie and Fjeldså 2005 in the Bibliography).

Description

- The Rubeho Forest Partridge is a small, approximately 29-cm-long (150 g) partridge, with orange-brown upper plumage boldly barred and spotted black, a rufous face, grey underparts and an olive-brown crown.
- They have a red bill, brown iris and yellow legs.
- Both sexes are similar.
- They are smaller than Udzungwa Forest Partridge, lack the white necklace and ochre-coloured vent, and have a speckled face, less distinctly barred secondaries, and scaly-patterned wing-coverts.
- Juveniles and downy chicks are undescribed.

Distribution and habitat

- They are endemic to Mbugu Hill in the Mafwemiro Forest, northern Rubeho Highlands, Dodoma region in southern Tanzania, where they are scarce.
- They inhabit and are restricted to mature montane forests 1 700–1 900 m above sea level with yellowwood (*Podocarpus*) trees and a thick understory.

Habits

They are shy and largely remain concealed in the undergrowth of the forest floor. They have a whistled call with high peeping notes.

Food and feeding

Although little is known about their diet it is probably similar to that of the Udzungwa Forest Partridge, consisting mainly of beetles, ants, flies, woodlice and seeds.

Breeding

There is currently no information available on their breeding biology.

Status and conservation

Because of ongoing habitat loss, their small population size, limited range and overhunting, the Rubeho Forest Partridge is considered to be Endangered (The IUCN Red List of Threatened Species v 2015-4), facing a high risk of extinction in the wild.

Madagascan Partridge
Caille de Madagascar
Margaroperdix madagarensis
Tetrao madagarensis Scolopi, 1786, Madagascar

Female Madagascan Partridge (photo by Werner Suter)

FAR LEFT Male Madagascan Partridge (photo by Markus Lagerqvist)

LEFT Male Madagascan Partridge (photo by Louise Jasper)

> ## Classification
> Madagascan Partridge are monotypic with no subspecies recognised.

Description

- Males have a reddish-brown crown with a diagnostic black-and-white mask and a black throat. The sides of the head and throat are dark grey, with a white stripe above the eye and along the sides of the throat.
- The upper breast is chestnut, while the lower breast and belly are black with distinct large oval white spots.
- The wings, flanks and tail are barred buff, chestnut and black.
- Females are tan-brown with black 'v' shaped barring on the back of the head and the throat down to the lower belly.
- The mantle, lower back and wings are dark brown with pale stripes similar to males.
- Juveniles are similar to but duller than females.
- Downy chicks are similar to those of *Coturnix* quails, being pale buffy-brown above, with dark brown lateral crown stripes and a broad dark brown stripe on the sides of the body, and dingy buff on the head with a blackish line below the eye.

Distribution and habitat

- They are an uncommon endemic across Madagascar, but absent from the extreme southern spiny forest.
- They are commonly seen near Perinet-Analamazaotra and between Toliara and Ihosy.
- They prefer grasslands, forest edges and forest clearings, weedy cultivated lands and heathlands in higher areas.

Habits

They are often found in pairs or small groups of 5–8 individuals, which run or fly to close cover when disturbed. They are mostly silent except for an occasional loud, repeated '*cou*' advertisement call and a short, throaty contact call.

Food and feeding

They forage while walking slowly, scratching the ground with their feet, picking at vegetable matter and occasionally lunging at insects. Their diet consists of seeds, berries, roots, green shoots and insects.

Breeding

Their breeding season is from March to June, possibly earlier as well. The nest is situated on the ground, well hidden in a tuft of grass or under a bush. The clutch is usually 8–15, possibly more, green-buff eggs with dark brown or black markings, and their incubation period is 18–19 days.

Status and conservation

Although they are not considered to be threatened, there is evidence of population declines, and they are sensitive to habitat degradation resulting from annual fires, which convert woody habitats to grassland. They are also heavily exploited through hunting with dogs, and are caught by trappers using rice-baited snares.

Stone Partridge
Poulette de rocher

Ptilopachus petrosus

Tetrao petrosus (Gmelin, JF, 1789), The Gambia

Stone Partridges in Kenya (both photos by Jacques Pitteloud)

Stone Partridges in Nigeria (photo by Tasso Leventis)

Stone Partridge in Nigeria (photo by Tasso Leventis)

Stone Partridge in Kenya (photo by Jacques Pitteloud)

Classification

The Stone Partridge is a sister species to Nahan's Partridge *P. nahani*, which in turn, are sister species to the New World quails. There are four subspecies recognised:

P. p. petrosus (Gmelin, JF, 1789) from The Gambia and Senegal to Cameroon

P. p. brehmi Neumann, 1908 in southern Chad to Sudan

P. p. major Neumann, 1908 in northern Ethiopia

P. p. florentiae Ogilvie-Grant, 1900 in South Sudan and southern Ethiopia to north-eastern Democratic Republic of the Congo, northern Uganda and central Kenya

Description

- The Stone Partridge is a small, plain, dark, bantam-like partridge weighing about 190 g with a stout body, long tail usually held cocked, short neck, red facial skin and relatively short dull red legs with no spurs.
- At close range they show barred flanks and a creamy-white belly patch, which is paler in the females.
- Juveniles are similar to adults, but with distinct barring on the back, underparts, rump, tail and inner secondaries.
- Downy chicks have a blackish chestnut forehead, crown and back, and are dark brown with black speckles below.

Distribution and habitat

- They are distributed in a broad band from The Gambia and Senegal across to Sudan, South Sudan, northern Uganda and north-western Kenya, with isolated populations in Ethiopia.
- A large part of their range is in the Sudanian savanna.
- They are locally common and prefer dry rocky hillside scrubland, steep dry river courses, lightly wooded rocky areas and sandy plains with scattered vegetation.

Habits

They have a characteristic behaviour of cocking and fanning their tail as they move. Usually seen in pairs or small coveys, they scamper over and around boulders between which they roost at night. They typically scatter and run from any disturbance, seldom flushing, but fly fast and direct if flushed. Small groups often call together, with some duetting at dawn and dusk. The call starts abruptly and is a continuous, high pitched, far-carrying piping or flute-like '*oo-wirr'oo-wirr'oo-wirr* …', rising and falling in intensity, while other individuals utter soft trilled churrs. They are not dependent on fresh water.

Food and feeding

They forage mostly during the cooler parts of the day, resting in shade during midday. Their diet consists of grass and herb seeds, fruits, green leaves and buds, and insects during the breeding season.

Breeding

They are probably monogamous and breed following the onset of the rainy season in dry areas and during the dry season in wet areas. The nest is a simple scrape, sometimes lined with grass and invariably well concealed among rocks or vegetation. The clutch of 4–6 eggs are pale ochre-yellow, slightly pointed at one end, and measure 33.3 x 24.8 mm.

Status and conservation

Although they are known to be hunted for food, there is no real concern about their status.

Nahan's Partridge
Francolin de Nahan

Ptilopachus nahani

Francolinus nahani Dubois, AJC, 1905, Popoie, Aruwimi River, Zaire (DRC)

Nahan's Partridges among dense forest undergrowth (photos by Robert Tizard)

Classification

Nahan's Partridge is a sister species to the Stone Partridge *P. petrosus*, which in turn are sister species to the New World quails, thus the change of English common name and genus and the assignment to the family Odontophoridae.

Description

- Nahan's Partridge is small with a mottled and streaked black-and-white belly, not barred, with bare red skin around the eye and a mostly white throat.
- Their legs are red with no spurs.
- Both sexes are very similar.
- Latham's Francolin *Peliperdix lathami* has a barred belly, no red skin on the face and yellow legs.
- Juvenile Nahan's Partridge are darker than adults with grey legs.
- Downy chicks are not described.

Distribution and habitat

- This monotypic partridge is localised in north-eastern Democratic Republic of the Congo, far western and southern-central Uganda.
- They are only found in the thick understory of dense primary forest, preferring riverine or swampy areas with drier banks, up to 1 400 m.

Habits

They are most often seen in pairs, or small groups up to 8 individuals, and cock their tail similar to that of the Stone Partridge. They are very shy, preferring to remain concealed within thick undergrowth of dense primary rainforest. Their call is a rising-falling '*keh keh*' wave, similar to that of the Stone Partridge, and most often heard in the early morning. Coveys also erupt into a series of rapid growls followed by complex whistles, which grow louder and then stop abruptly.

Food and feeding

They forage by scratching on the forest floor in leaf litter. Their diet consists of seeds, bulbs, green shoots, insects and small molluscs.

Breeding

They are probably monogamous and possibly breed throughout the year. They nest in unlogged primary forest. One record of a nest was located in the hollow of a tree trunk about 1 m off the ground. The four eggs in that clutch were smooth, glossy buff to pinkish, with pale brown and purple speckles, and measured 36.0 x 26.0 mm.

Status and conservation

They are more rare than Latham's Francolin and they may be vulnerable to hunting pressure and forest degradation due to timber extraction. Their conservation status is therefore listed as Endangered (The IUCN Red List of Threatened Species v 2015-4).

PART 3
Spurfowls

SPURFOWLS

Classification

- The 24 African spurfowl species are all placed in one genus *Pternistis* Wagler, 1832.
- They can be grouped into four groups according to diagnostic plumage or habitat preference features: bare-throated, montane, scaly or vermiculated (see Table 2).

Table 2: Recognised spurfowl species groups (*Pternistis* spp.)

Bare-throated	Scaly
Red-necked *P. afer*	Ahanta *P. ahantensis*
Swainson's *P. swainsonii*	Scaly *P. squamatus*
Yellow-necked *P. leucoscepus*	Grey-striped *P. griseostriatus*
Grey-breasted *P. rufopictus*	
Montane	**Vermiculated**
Erckel's *P. erckelii*	Double-spurred *P. bicalcaratus*
Djibouti *P. ochropectus*	Heuglin's *P. icterorhynchus*
Chestnut-naped *P. castaneicollis*	Clapperton's *P. claffertoni*
Black-fronted *P. atrifrons*	Hilderbrandt's *P. hilderbrandti*
Jackson's *P. jacksoni*	Natal *P. natalensis*
Handsome *P. nobilis*	Hartlaub's *P. hartlaubi*
Mount Cameroon *P. camerunensis*	Harwood's *P. harwoodi*
Swierstra's *P. swierstrai*	Red-billed *P. adspersus*
	Cape *P. capensis*

Description

- Spurfowls are generally larger than francolins and weigh 340–950 g.
- They have black, red or orange legs, and the adult males have long, sharp leg spurs, generally two spurs on each lower leg, one above the other.
- They have streaked or vermiculated back feathers.

Natural history

Their distribution is restricted to sub-Saharan Africa, with only Double-spurred Spurfowl *P. bicalcaratus* extending into Morocco, North Africa, and no species occurring in Asia. They live in bushy and wooded habitats, regularly take refuge or sometimes roost in trees, and prefer to run, rather than flush, to escape danger. Courting males will typically circle and approach a female with wings bowed and feathers puffed to induce her to pair with him. The hen is protective of her brood and will confront any threat with wings spread and raised. Their advertisement calls are raucous and grating or cackling, much more like those of pheasants and chickens, and they tend not to react to tape recordings. Their chicks have a more extensive crown patch than those of the francolins and only one broad eye-stripe.

Evolutionary placement

- As mentioned in the introduction to the francolins and partridges, spurfowls (and francolins) were previously classified with partridges in the tribe Perdicini within the family Phasianidae, although the only anatomical feature that supports this grouping unambiguously is that they both have 14 tail feathers.
- Until recently, all 41 spurfowl and francolin species (36 African, five Asiatic) were lumped into a single genus, *Francolinus*, but divided among seven groups, at least one member of which may be found in all habitats throughout sub-Saharan Africa outside of true desert.
- However, many birders, farmers and wing-shooters divided these groups into two assemblages:
  - francolins (of the vermiculated and bare-throated spurfowl groups); and
  - partridges (making up the red-tailed, red-winged and striated francolin groups).
- Scientifically, there is no problem in naming them because the generic name *Pternistis* (Greek for 'one who trips with the heel', referring to double spurs of males) has frequently been used for many spurfowls and is therefore available.
- Since 'partridges' is a taxonomic mosaic, 'partridge' is probably not an acceptable common name for the quail-like francolins.

Indeed, even 'the' Partridge *Perdix perdix* is grouped with the 'true' pheasants (for example, *Phasianus* species), family Phasianidae.

The common name 'spurfowls' is therefore supported for this group because it has already been used for the Yellow-necked Spurfowl *Pternistis leuscoscepus* of East Africa, which is a member of the bare-throated group and is a close relative of the Swainson's Spurfowl *P. swainsonii* and the Red-necked Spurfowl *P. afer*. Indeed, there is evidence that these species can interbreed.

Conservation

Another important reason to have separate common names for spurfowls and francolins is the fact that, at least some populations of various spurfowls can withstand sustainable utilisation, whereas with the exception of the Grey-winged Francolin *Scleroptila afra*, African francolins tend to occur at densities that cannot withstand sustained wing-shooting. In some regions, habitat structure and forage plants are managed to sustain or increase spurfowl numbers for sustained wing-shooting.

Only four of the 24 species covered in this section are threatened; Djibouti Spurfowl *P. ochropectus* is listed as Critically Endangered, Mount Cameroon Spurfowl *P. camerunensis* as Endangered, and Swainson's Spurfowl and Harwood's Spurfowl *P. harwoodi* as Vulnerable (The IUCN Red List of Threatened Species v 2015-4).

Red-necked Spurfowl
Francolin à gorge rouge

Pternistis afer

Tetrao afer (Statius Müller, PL, 1776), Benguella, Angola

Red-necked Spurfowl subspecies
P. a. humboldtii from Gorongoza, Mozambique
(photo by Maans Booysen)

Red-necked Spurfowl subspecies
P. a. cranchii from Lake Mburo National Park,
Uganda (photo by Cliff Dorse)

Red-necked Spurfowl subspecies *P. a. afer*
from the Kunene River, northern Namibia
(photo by Ian White)

Red-necked Spurfowl from Caia, Mozambique (photo by Maans Booysen)

Red-necked Spurfowl from Mikumi National Park, Tanzania (photo by Ian White)

Red-necked Spurfowl from Tanzania (photo by Tasso Leventis)

Red-necked Spurfowl from Kenya (photo by Jacques Pitteloud)

Classification

They are most closely related to Grey-breasted Spurfowl *P. rufopictus*, Yellow-necked Spurfowl *P. leucoscepus* and Swainson's Spurfowl *P. swainsonii*, which hybridise readily. Such a hybrid was incorrectly described as a new 'species' *P. cooperi* during the 1940s. With widespread geographical variation, as many as 20 subspecies have previously been described, but the latest suggestion is that three subspecies should be recognised:

P. a. cranchii (Leach, 1818) in western Congo eastward to East Africa and south to northern and eastern Angola, north-eastern Zambia, northern Malawi and western Tanzania

P. a. afer (Statius Müller, PL, 1776) in Angola and northern Namibia and disjunctly along the southern coast of South Africa, the Limpopo Province and the Mpumalanga Lowveld

P. a. humboldtii (Peters, 1854) from eastern South Africa and Zimbabwe through Mozambique, north-eastern Zambia to Kenya

These latter two subspecies may be recognised as one full species distinct from *P. cranchii* in the future.

Description

- They have bare red skin on the throat and around the eyes, as well as red legs and bill.
- There is considerable colour variation, from brown to almost black.
- On the crown and neck, the feathers may be brown with black streaks, or black with white, elongated markings.
- The throat feathers are mainly white, with chestnut and black markings along the shaft and on the outer edge.
- The breast and flank feathers are dark brown to black, with single narrow white longitudinal streaks on either side of the shaft.
- Males have long spurs, sometimes two on each leg.
- Juveniles resemble adults but have dull browner plumage and a white throat.
- The underparts are barred faintly with black and white.
- Downy chicks are rufous-brown above, with dark brown stripes on the crown and back.
- The sides of the head are buff, with a black stripe through the eye and extending downwards to the breast.
- The throat is buff, grading to yellowish-buff on the belly, flecked with dull black.

Distribution and habitat

- They occur from the central African forests and Kenya in the north through Zambia, Malawi and Tanzania down to eastern Zimbabwe, Mozambique and South Africa.
- They prefer dense cover and moist habitats, most frequently occurring on the edges of evergreen forests, or in riparian thickets or marsh edges.

Habits

They occur in pairs or small groups of 3–6 individuals. They are shy and reluctant to fly when disturbed, preferring to run into dense cover. When flushed, they fly fast and normally perch in a tree. At night, they roost in trees. Although territorial and apparently seasonally monogamous, males that have an incubating hen will court other females that enter his territory. Males do not assist with care of their offspring. They are easily overlooked and their harsh '*krr krr krr korWA, korWA, korWA*' call, usually uttered at dawn and dusk, is similar to but more regular in volume than that of Swainson's Spurfowl.

Food and feeding

The diet consists primarily of small tubers, bulbs, roots, seeds, fruits and other vegetable material during the winter, and insects, termites and molluscs such as snails during the summer. They will also eat waste commercial grains such as maize, sorghum and cereals.

Breeding

Their breeding season is variable from the late austral summer in South Africa, during the wet and dry seasons, January–July (mainly April) and November–December (mainly December), in Zimbabwe, February–June in Angola, November–August (mainly February–May) in Zambia and January–September (mainly February–June) in Malawi. The nest, lined with grass and feathers, is constructed by the female, and is often placed in dense grass at the base of a tree or bush. The clutch of 3–7, rounded to oval creamy white or creamy brown thick-shelled eggs measure 45.0 x 35.9 mm. Incubation lasts 23 days. Chicks can fly at 10 days and are almost fully grown at 3–4 months.

Status and conservation

They are impacted by the degradation and removal of forest and woodland edge by burning, heavy grazing and extensive cultivation. They also respond to droughts by moving from a dry area to more moist sites.

Swainson's Spurfowl
Francolin de Swainson

Pternistis swainsonii

Perdix swainsonii (Smith, A, 1836), Kurrichane, (Zeerust) North West Province, South Africa

Male Swainson's Spurfowl on a conspicuous territorial perch (photo by Ian White)

ABOVE Male Swainson's Spurfowl calling from a prominent perch (photo by Andre Botha)

BOTTOM RIGHT Female Swainson's Spurfowl (photo by Ian White)

Female Swainson's Spurfowl with two juvenile chicks (photo by Hugh Chittenden)

Male Swainson's Spurfowl (photo by Ian White)

Male Swainson's Spurfowl (photo by Andre Botha)

Calling Swainson's Spurfowl (photo by Hugh Chittenden)

Classification

They are most closely related to Yellow-necked Spurfowl *P. leucoscepus*, Red-necked Spurfowl *P. afer* and Grey-breasted Spurfowl *P. rufopictus*. A new 'species', *P. cooperi*, described during the 1940s was later correctly identified as a hybrid with Red-necked Spurfowl. Two subspecies are recognised:

P. s. lundazi White, CMN, 1947 in north-western Zimbabwe to southern Mozambique

P. s. swainsonii (Smith, A, 1836) in south-eastern Angola, northern Namibia, north and eastern Botswana, and South Africa

Description

- They are sexually dimorphic in size, dark brown overall, and have bare red skin on the throat and around the eyes, but have black legs and upper bill.
- They also lack the distinct white or grey patterning on the breast and flanks.
- Males have large black leg spurs.
- Juveniles resemble the adults, but are paler and duller overall, and the throat has white feathers.
- The belly is white, finely barred with black.
- Downy chicks are dark brown above, with darker streaks on the crown and back.
- The sides of the head are buffy, with a single black eye-stripe.

Distribution and habitat

- They occur in southern Africa, extending into southern and eastern Zambia.
- They have expanded their range into maize-growing areas in KwaZulu-Natal, Mpumalanga, Lesotho and parts of Zimbabwe, where they exploit crops and the associated weeds and introduced alien trees.
- At least part of this range expansion, particularly into KwaZulu-Natal, the Limpopo Province and Eastern Cape, was assisted by deliberate introductions.
- They are common in most savannas and tall grasslands adjacent to cultivation and water, except in the arid western South Africa and Namibia.

Habits

They occur in pairs and small groups. Although bold, they are wary as a rule. During the morning and late afternoon, they venture from cover into open ground to forage and dust-bathe. In the heat of the day they rest in the shade of scrub and brush. They feed at night in open lands during full moon, particularly following hot days. Although fond of cultivated lands, they favour weedy land edges and contour strips when these are adjacent to tall grass and trees. They also favour river banks and rank growth of sponges and marsh edges. When disturbed, they run into cover or flush and fly to the closest cover. A flushed bird will fly into grass or scrub cover from which it is difficult to re-flush, or will perch in a tree. They roost in low trees and bushes, and sometimes remain perched on wet mornings. The call is a harsh '*krrraa, krrraa, krrraa*', sometimes rendered as '*kwahli, kwahli*', repeated several times and descending in pitch and volume towards the end. It is similar to that of Red-necked Spurfowl, whose call is more even in pitch and volume. Calling is most frequent at dawn and dusk, and calling males perch conspicuously on a termite mound, fence post or tree stump. They also have an extended high-pitched mewing contact call.

Food and feeding

The winter diet consists of the seeds of grass and agricultural weeds, bulbs, tubers, roots and berries. In summer they eat termites, worms, beetles, grasshoppers, ticks, spiders and small molluscs. They also eat waste grains such as maize, sorghum, barley and cereals. They are accused of digging out crop seedlings, although the amount of pioneer weed and grass seeds and seedlings consumed, along with the high intake of worms ingested, make up for any perceived damage that they may cause. Prior to roosting, groups move towards water to drink. However, water does not limit their distribution, with populations flourishing in semi-arid environments where watering points are few and well out of reach of the majority of birds.

Breeding

Their extended breeding season peaks from February to May. They are monogamous and territorial while breeding. Their nest is a shallow bowl lined with grass and leaves, usually concealed within dense cover of weeds or shrubby vegetation. They will also nest among crops or in old lands. The clutch of 4–6, sometimes up to 12, rounded to oval pinkish-cream or buff eggs measure 43.9 x 35.7 mm. Incubation, by the female only, starts after the clutch is complete and lasts 21–24 days. Chicks can flutter-fly at 14 days and are almost adult size at 3 months.

Status and conservation

They are tolerant of human-transformed habitats, increasing population sizes to levels above those of natural populations in parts of Zimbabwe, Gauteng, Limpopo Province, KwaZulu-Natal and Northern Cape, where they can support commercial wing-shooting. Habitats are often modified, by adding water points and plants suitable for food and cover, especially along the interface between croplands and natural habitats, to encourage the growth of their populations. However, repeated hunting on the same property within one season can reduce populations to levels that cannot be hunted sustainably. Hunting of this species should also be delayed until August to allow for completed breeding and full development of the chicks.

Yellow-necked Spurfowl
Francolin à cou jaune
Pternistis leucoscepus
Francolinus leucoscepus (Gray, GR, 1867) Ethiopia

Yellow-necked Spurfowl
(photo by Lorenzo Barelli)

Yellow-necked Spurfowl from Kenya
(photo by Tasso Leventis)

Yellow-necked Spurfowl from Kenya
(photo by Tasso Leventis)

GAMEBIRDS

Female Yellow-necked Spurfowl with a brood of chicks from Tarangire National Park, Tanzania (photo by Ron Eggert)

Female Yellow-necked Spurfowl from Kenya (photo by Jacques Pitteloud)

Female Yellow-necked Spurfowl from Tarangire National Park, Tanzania (photo by Sergey Dereliev)

Classification

The Yellow-necked Spurfowl is most closely related to the Swainson's Spurfowl *P. swainsonii*, Red-necked Spurfowl *P. afer* and Grey-breasted Spurfowl *P. rufopictus*. They are known to hybridise with Grey-breasted Spurfowl. They are monotypic with no subspecies recognised, although two subspecies are recommended:

P. l. leuoscepus (Gray, 1867) and *P. l. infuscatus* Cabanis, 1868

Other putative subspecies (*holtemülleri*, *keniensis*, *kilimensis*, *tokora*, *muhamed-ben-abdullah*) should be synonymised with *P. l. infuscatus*.

Description

- They are dark brown with a characteristic bright yellow bare skin throat, red face and black bill and legs.
- In flight they have a conspicuous buff patch in the wing.
- Juveniles are less well marked with a dull yellow throat.
- Downy chicks have a dark brown forehead, a light brown crown with a dark brown crown stripe and a single eye stripe from behind the eye to the nape.
- Their back is dark brown with two buff and brown stripe on each side.

Distribution and habitat

- They are distributed across southern and northern Somalia, eastern Ethiopia, Eritrea, south-eastern South Sudan, far north-eastern Uganda, most of Kenya and northern Tanzania.
- Their preferred habitats are open savanna, scrub and adjacent cultivated lands.
- They generally avoid tall grass.

Habits

They are usually seen in pairs or small groups, but will congregate at sites with good food availability. They roost in low trees, emerge to forage and dust-bathe during the early morning, and rest in scrub shade for most of the rest of the day. When disturbed, they prefer to run into cover, but when flushed will often fly to perch in a tree. Their advertisement call, similar to that of Swainson's Spurfowl, is often given from a conspicuous elevated perch. The call is a repeated raucous descending '*kerak*' or '*ko-warrrk*', mostly at dawn and dusk.

Food and feeding

Most foraging is during the early morning. Their diet consists mostly of corms and tubers of sedges *Cyperus rotundus*, the fruits and seeds of herbs and grasses, plant parts and insects, mostly termites. They will also eat waist grains in cultivated lands. A common foraging behaviour is to scratch through rhinoceros (Rhinocerotidae) and African Elephant *Loxodonata africana* dung for undigested seeds, insects and their larvae.

Breeding

They are monogamous, breeding at any time of the year, but mostly during the first half of the year. The nest is a scrape in the ground, usually unlined. The clutch of 3–8, usually 5, eggs are oval with a slight point and are cream-to-pale buff with darker speckles and white pore marks, and measure 45.2 x 35.5 mm. Incubation is for 18–20 days, and chicks can fly short distances at 2 weeks.

Status and conservation

Although previously abundant throughout most of their range, numbers have declined dramatically where human populations have increased. Noose snaring is the main culprit for hunting of this species, particularly along fence lines adjacent to cultivated lands.

Grey-breasted Spurfowl
Francolin à poitrine grise

Pternistis rufopictus

Pternistis rufopictus Reichenow, 1887, Wembere Steppe, Tanzania

ABOVE AND LEFT Male Grey-breasted Spurfowls from Serengeti National Park, Tanzania (photos by Adam Riley above & Dalton Gibbs left)

Classification

They are most closely related to Red-necked Spurfowl *P. afer*, followed by Yellow-necked Spurfowl *P. leucoscepus* with which they are reported to hybridise, and Swainson's Spurfowl *P. swainsonii*. Their evolutionary origin may indeed be a hybrid between the various taxa of the bare-throated spurfowl group. They are monotypic with no subspecies recognised.

Description

- The Grey-breasted Spurfowl is a large dark brown spurfowl (average male mass 848 g, average female mass 588 g) with broad chestnut streaks on their back.
- They differ from Yellow-necked and Red-necked Spurfowls by having an orange bill, grey-brown legs, orange-pink skin on the throat, and white moustachial stripes.
- In flight, they show pale bases to the primary feathers.
- Juveniles are similar to adults but their upperparts are grey-black with white central shaft streaks and barring, and grey-buff margins.
- Downy chicks have a dark crown with rufous-brown margins and two dark brown eye-stripes above and below the eye.

Distribution and habitat

They are endemic to the plains and savannas of north-western Tanzania. They are fairly common in grasslands, Umbrella Thorn *Vachellia tortilis* scrub and woodlands, and riverine thickets from the southern Serengeti west to Mwanza.

Habits

They usually occur in pairs and small coveys, but may congregate into comparatively large groups in favoured feeding areas. They often venture into open grasslands at dawn and dusk. They call mostly at dawn and dusk, usually from an elevated perch. Their crowing call '*ka-waaaark, ka-waaaark, ka-waarrrk*' or '*koarrrk-koarrrk-karrkkrrk-krrk-krr*' descending towards the end is similar to that of Yellow-necked Spurfowl, but is faster. Their alarm call when flushed is a high-pitched cackle.

Food and feeding

Their diet consists mainly of sedge *Cyperus* spp. corms, tubers, grass and weed seeds and insects, including grasshoppers and termites. They will also eat waste cereal grain and legume seeds in cultivated lands.

Breeding

They are monogamous, breeding mostly from February to April, but can be as late as July. The nest is a scrape between long grass, often lined with grass and feathers. The clutch of 4–5 buff or pale brown eggs with chalky white pore spots measure 43.4 x 35.2 mm.

Status and conservation

Although still relatively common, they have been impacted by habitat degradation and transformation by extensive agriculture and overgrazing.

Erckel's Spurfowl
Francolin d'Erckel

Pternistis erckelii

Perdix erckelii (Rüppell, 1835), Taranta Mountains, Ethiopia

ABOVE Female Erckel's Spurfowl (photo by Marvin Hyett)

RIGHT Male Erckel's Spurfowl (photo by Dick Daniels)

Classification

They are most closely related to Djibouti Spurfowl *P. ochropectus*. They are a monotypic spurfowl with no subspecies recognised.

Description

- The Erckel's Spurfowl is a large, grey-and-chestnut-striped spurfowl with a striking head pattern, including a black forehead and supercilium, a chestnut crown, grey ear coverts and a white throat.
- The lower neck is grey like the upper belly, but has greyish brown margins and a thin central buff streak, whereas the upper belly feathers have central greyish-black streaks.
- The lower belly feathers have a broad buff central streak constricted in the middle and expanded distally into a tear drop, margined with rufous.
- They have no bare facial skin and have a black bill and yellowish legs.
- They are more olive-brown on top than Chestnut-naped, Black-fronted and Djibouti Spurfowls.
- Females are similar but smaller than males.
- Juveniles are paler grey on top with quail-like streaks and bars on the back.
- Downy chicks have a dark stripe from the forehead down the centre of the crown.
- The sides of the head have a buff-and-black moustachial stripe, through the eye down the side of the neck.
- The upper body is dark brown and black with a buff lateral stripe on each side.

Distribution and habitat

- They occur in the highlands of central and northern Ethiopia and Eritria with a small population in the Red Sea province of north-eastern Sudan.
- They are locally common above 2 000 m above sea level in giant heath *Erica arborea*, on scrub-covered hill slopes and cliffs, and in forest edges and remnants.

Habits

They are shy and sedentary, mostly occurring in pairs or small groups of up to 8 individuals. When disturbed they run for cover or flush and fly downhill. They forage during the early morning within scrub and forest edges, sometimes entering cultivated lands to feed on waste grain. They are generally quiet, calling occasionally from a conspicuous perch at dawn and dusk. Their call is an extended harsh cackling crow 'errk-erkk-erk-erk-rkkuk-kuk-ku', ending in a chuckle-like rattle.

Food and feeding

They also forage on sheer cliffs by flying from ledge to ledge. Their diet consists of grass seeds and shoots, berries and seeds of shrubs and herbs, and some fallen grain and insects.

Breeding

They are monogamous, breeding mostly during the rainy months of April and May, and from September to November. Their nest is a typical scrape on the ground. The clutch of 4–10 hard-shelled dirty white to pale brown eggs measure 46.2 x 36.5 mm.

Status and conservation

They are heavily impacted by widespread deforestation, particularly in Ethiopia.

Djibouti Spurfowl
Francolin des Somalis

Pternistis ochropectus

Francolinus ochropectus (Dorst & Jouanin, 1952), Plateau du Day, near Tadjoura, Djibouti

Djibouti Spurfowl
(photo by Callan Cohen)

Djibouti Spurfowls
(photos by Nik Burrow)

Classification

They are most closely related to Erckel's Spurfowl *P. erckelii*, followed by Chestnut-naped Spurfowl *P. castaneicollis* and Black-fronted Spurfowl *P. atrifrons*. It is a monotypic species with no subspecies recognised.

Description

- The Djibouti Spurfowl, also sometimes known as the Ochre-breasted Spurfowl, resembles a dark drab Erckel's Spurfowl, but they have a black face with a short white supercilium behind the eye, and their ranges do not overlap.
- They are also similar in appearance to Chestnut-naped Spurfowl and Black-fronted Spurfowl.
- Females are similar to males with a slightly more rufous tail.
- Juveniles are similar to adults with more buff-and-grey barring than streaking.
- Downy chicks are undescribed.

Distribution and habitat

- They are endemic to the dense evergreen forests of Djibouti where they are confined to a small area around Forêt du Day and the nearby Mabla Mountains, between 700 and 1 800 m above sea level.
- They are mostly found in the undergrowth of juniper forests *Juniperus procera*, dry secondary woodlands and river courses.

Habits

They occur in pairs and family groups that forage during the early morning and spend much of the day roosting in trees. They are shy and more often heard than seen. Their call is a loud accelerating '*Erk-erk-ka-ka-krrrr*' rattle, similar to that of Erckel's Spurfowl. Foraging individuals also have a soft contact clucking.

Food and feeding

When foraging, they search for food by raking over the ground, including in areas which have been disturbed by warthogs *Phacochoerus africanus*. Their diet consists of berries, grass and other seeds, figs and termites.

Breeding

They breed in dense, lush vegetation from December to February. The only nest discovered to date was a shallow depression lined with grass on a mountain ledge. Their clutch is 5–7 eggs. Little more is known about their breeding biology.

Status and conservation

They are Critically Endangered (The IUCN Red List of Threatened Species v 2015-4), and listed on the CITES Appendix II, although Djibouti is not a signatory nation. With fewer than 500 individuals estimated in the population, they are threatened by habitat destruction due to deforestation for firewood, overgrazing by domestic livestock, including cattle, camels and goats, and possible climate change, droughts and fungal disease.

Chestnut-naped Spurfowl
Francolin à cou roux

Pternistis castaneicollis

Francolinus castoneicollis (Salvadori, 1888), Lake Ciar-Ciar (= Chercher), Shoa (Hararghe), Ethiopia

ABOVE Female Chestnut-naped Spurfowl (photo by Nik Borrow)

LEFT Male Chestnut-naped Spurfowl (photo by Otto Schmidt)

Classification

Black-fronted Spurfowl *P. atrifrons* was recently split from Chestnut-naped Spurfowl, and they together form the sister group to a clade comprising Erckel's Spurfowl *P. erckelii* and Djibouti Spurfowl *P. ochropectus*. Three subspecies are recognised:

P. c. ogoensis (Mackworth-Praed, 1920) in northern Somalia

P. c. castaneicollis (Salvadori, 1888) in west-central to eastern Ethiopia

P. c. kaffanus (Grant, CHB & Mackworth-Praed, 1934) in south-western Ethiopia

Description

- The Chestnut-naped Spurfowl is a large richly coloured spurfowl with conspicuous chestnut-coloured flanks and at least some chestnut neck feathers.
- They are the same size as Erckel's Spurfowl but are more brightly coloured with a chestnut head and neck, and bright red bill and legs.
- They lack the bare red face of the Harwood's Spurfowl *P. harwoodi* and differ further by having an overall rusty neck, breast and cap.
- Black-fronted Spurfowl *P. atrifrons*, confined to the mega-region of southern Ethiopia, are smaller and paler with no chestnut on the flanks or neck and have a diagnostic single bald yellow skin patch just above the ear.
- Females are similar to males but are smaller.
- Juveniles are duller, have a black-and-buff barred and vermiculated rump and tail.
- Downy chicks are undescribed.

Distribution and habitat

- They are common to abundant in northern Somalia through south-central Ethiopia.
- Their preferred habitats are in montane heath moorlands, juniper forests, forest glades, and edge and scrubby areas at lower altitudes.

Habits

They generally occur in small groups or pairs and emerge from cover to feed, particularly during wet weather. When disturbed they will flush in open cover, otherwise they will retreat to dense cover where they will remain. Where vegetation is tall enough they will roost on elevated perches, but will also roost on the ground. They call from dense cover, mostly during the early morning and evening, but will call throughout the day. Their call is a noisy '*kek kek kek keraak*' crowing, often antiphonal between two individuals.

Food and feeding

They feed throughout the day, particularly during wet weather. Their diet consists of seeds, insects and termites.

Breeding

They are monogamous and breed mostly during the drier months, from January to March. The nest is a scrape under a bush or within cover. The clutch of 5–6 smooth round cream-coloured eggs measure 46.8 x 36.8 mm. Their incubation biology is unknown.

Status and conservation

Although they are easily snared in dense cover, they are relatively tame in some areas and are not heavily utilised.

Black-fronted Spurfowl
Francolin à front noir

Pternistis atrifrons

Pternistis (*castaneicollis*) *atrifrons* (Conover, 1930), Boran (Mega), Ethiopia

ABOVE Male Black-fronted Spurfowl (photo by Kai Gedeon)

LEFT Black-fronted Spurfowl showing the absence of any chestnut on the flank and neck feathers (photo by Kai Gedeon)

Classification

- During an expedition in 1929 to Ethiopia, a 'francolin' was collected from the Mega Mountains for the Field Museum Chicago which was described as a new species, the Black-fronted (or Ethiopian) Francolin.
- Although later treated as a subspecies of the Chestnut-naped Francolin (Spurfowl) *P. castaneicollis*, this was regularly challenged.
- Recently, expeditions to this area during 2012 and 2013 by Till Töpfer and others (see Töpfer et al. in the Bibliography) concluded that the separate taxonomic status of *P. atrifrons* is warranted. Most of the information in this species account is gleaned from their findings.
- With Chesnut-naped Spurfowl, they form the sister group to a clade comprising Erckel's Spurfowl *P. erckelii* and Djibouti Spurfowl *P. ochropectus*.

Description

- They are smaller and paler than Chestnut-naped Spurfowl with no chestnut-coloured flanks or chestnut neck feathers, and lack the red face of Clapperton's Spurfowl *P. clappertoni*.
- Males have bold black-and-buff scaly upperparts and a nearly plain off-white throat, breast and underparts.
- The characteristic black forehead and supercilium is separated by an ill-defined white line from the greyish-brown crown.
- The bill is coral-red and the legs orange-red with two equal-sized spurs.
- The eye-rings are yellowish with a diagnostic single bald yellow skin-patch just above the ear.
- Females are slightly smaller with no leg spurs, and have a smaller head with a weaker bill, much less black forehead and lack the pronounced black supercilium.
- Juveniles have a more streaky appearance with prominent whitish shaft streaks on the feathers of the lower neck and upper back and a bluish skin-patch above the ear, bluish eye-ring and orange legs.

Distribution and habitat

- This species is restricted to the moist mountainous areas of the Mega region in southern Ethiopia, and far northern Kenya.
- They occur in semi-open woodland with extensive low shrubs, interspersed by taller trees and bushes and open rocky and grazed patches between 1 480 and 2 223 m above sea level.

- Although previously reported to inhabit four different forest habitats (*Hagenia* forests, Highland bamboo *Arundinaria*, Juniper-*Podocarpus* forest, and Olive-*Podocarpus*-Juniper forest), many of these habitats have been severely transformed, such that they are now largely found in scrubby woodland where juniper trees are largely missing.

Habits

They generally occur in small groups or pairs which are shy and tend to hide in low vegetation when approached. They prefer to escape on foot and fly only relatively short distances before landing on the ground, not in trees or bushes. Their advertisement call, similar to that of the Chestnut-naped Spurfowl, is a harshly grating and fairly noisy cackle lasting about three seconds and consisting of four elements with pauses of about two seconds. This call is usually uttered from the ground, mostly at dawn and later during the day.

Food and feeding

Their diet is similar to that of other spurfowl inhabiting scrubby habits in mountainous terrain.

Breeding

The breeding behaviour is probably very similar to that of Chestnut-naped Spurfowl. Evidence suggests that the timing of breeding is closely aligned with the primary rainy season from October to March. The nest is a scrape concealed under bush or in undergrowth, and the clutch is probably 5–6 eggs.

Status and conservation

Their range may be smaller and more fragmented than previously thought, making localised populations highly threatened. Their habitats may already be degraded, resulting mainly from grazing pressure, agricultural expansion, commercial firewood and timber exploitation. Hunting, largely by young people using nooses for trapping, appears to be the biggest direct threat to these isolated populations. They should probably be considered an endangered range-restricted species like the Djibouti Spurfowl.

Jackson's Spurfowl
Francolin de Jackson

Pternistis jacksoni

Francolinus jacksoni (Ogilvie-Grant, 1891), Kikuyu, Kenya

Jackson's Spurfowl (photo by Lorenzo Barelli)

Juvenile Jackson's Spurfowl (photo by Jacques Pitteloud)

Classification

Although considered with the montane spurfowl group, they are most closely related to the Grey-striped Spurfowl *P. griseostriatus* of the scaly group. They are a monotypic spurfowl with no subspecies recognised.

Description

- The Jackson's Spurfowl is a large rufous spurfowl with red legs and bill, and narrow red eye-rings.
- Males weigh over 1 kg.
- They have a buff throat and greyish lower neck with the upper part of the lower neck being similarly patterned to the rest of the belly.
- The lower neck feathers are chestnut-coloured edged with buffish to white, but the degree of chestnut and buff-and-white varies.
- They have white-streaked chestnut underparts.
- They are larger and more rufous than the Scaly Spurfowl *P. squamatus*.
- Females are similar to males but smaller.
- Juveniles are duller, with dark barring above and on the belly.
- Downy chicks are undescribed.

Distribution and habitat

- They are endemic to the montane forests and Afro-alpine zone of the western and central Kenyan mountains, especially the Aberdare range, as well as marginally into Uganda.
- They are locally common in montane moorlands, bamboo patches, *Podocarpus* and *Juniperus* forests and montane heath, from 2 200 to 3 700 m above sea level.
- They usually prefer dense shrubby thickets and moribund bamboo stands rather than primary forests.

Habits

They are particularly sedentary, usually confined to dense vegetation where they are more often heard than seen. However, they do often emerge onto short grassy patches and dust-bathe on sandy roads, particularly during the early morning and during wet weather. Where they have become accustomed to humans, such as at picnic places and lodges, they do become tame and confiding. Their call is a raucous high-pitched deliberate '*kirr-kee-kik*' crowing at dawn and dusk, similar to that of Scaly Spurfowl and Hildebrandt's Spurfowl *P. hildebrandti*. They also utter a low clucking contact call while foraging in dense vegetation.

Food and feeding

Their diet consists of bulbs, tubers, bamboo seeds, grass shoots, berries, insects and small snails.

Breeding

They are monogamous and breed during the drier months of the year. Little is known about their breeding biology, but their clutch size is 3–7 eggs, which are glossy pale brown and measure 46.5 x 36.0 mm.

Status and conservation

Although they are abundant in some national parks, they are dependent on park management for their future stability. They are also vulnerable to habitat degradation in lower forest edges throughout their range.

Handsome Spurfowl
Francolin noble

Pternistis nobilis

Francolinus nobilis (Reichenow, 1908), Virunga Volcanoes, Zaire

LEFT Handsome Spurfowl from the Bwindi Impenetrable Forest, Uganda (photo by Nik Borrow)

BELOW A pair of Handsome Spurfowl from Rwanda (photo by Tasso Leventis)

Classification

They are most closely related to Mount Cameroon Spurfowl *P. camerunensis*. They are a monotypic spurfowl with no subspecies recognised.

Description

- The Handsome Spurfowl is a large spurfowl (average male mass 877 g and female mass 624 g).
- They are dark reddish-brown, particularly in the wings, back and underparts, with a greyish-brown head, primaries and rump.
- Their wing coverts and lower neck are deep maroon, with light grey scalloping on the lower neck.
- Their throat is buff-white with the rest of the belly being chestnut with grey edges varying to broad buff or white edges.
- They have a bare red eye patch, as well as a red bill and legs.
- Females are smaller and duller than males.
- Juveniles are similar to adults but the upperparts are barred with dark grey and rufous-buff, while the underparts are paler.
- Downy chicks are undescribed.

Distribution and habitat

- They are endemic to the highlands of the Albertine Rift of the far western mountain forests of the Democratic Republic of the Congo, south-western Uganda, and borders between Rwanda and Burundi.
- They are locally common in dense undergrowth from the lower edge of the montane forests, through the bamboo thicket zone, up to the Afro-alpine heath zone, especially near water or in swampy areas, mainly at altitudes from 1 850 to 3 700 m above sea level.

Habits

They are sedentary and most easily seen in small groups or pairs, in early mornings and at dusk, in clearings and on tracks. They are shy and elusive, more often heard than seen, and when disturbed will run through thick undergrowth. If flushed, they fly a short distance before dropping back into cover. They are very vocal, particularly at dusk before roosting. Their call is a noisy musical crowing with a repeated '*chuk ker-ack*' or squealing '*ker-ack*' at dawn and dusk.

Food and feeding

They forage along roads in the early morning and late afternoon. Their diet consists largely of seeds.

Breeding

They are known to breed from May to September, but little is known about their breeding biology.

Status and conservation

Although common to locally abundant, they are trapped with snares for food by local people. They are locally dependent on bamboo, which is prone to destruction by heavy harvesting.

Mount Cameroon Spurfowl
Francolin du Mont Cameroun

Pternistis camerunensis

Francolinus camerunensis (Alexander, 1909), Cameroon Mountain, Cameroon

RIGHT Male Mount Cameroon Spurfowl (photo by Hadoram Shirihai)

BELOW LEFT Incubating hen Mount Cameroon Spurfowl (photo by Ashute Essebi Marcel)

BELOW Mount Cameroon Spurfowl nest and clutch (photo by Ashute Essebi Marcel)

Classification

The Mount Cameroon Spurfowl is most closely related to the Handsome Spurfowl *P. nobilis*. They are a monotypic spurfowl with no subspecies recognised.

Description

- They are a dark brown spurfowl with a red bill, facial eye patch and legs.
- They are darker than the similar Scaly Spurfowl *P. squamatus*, with red facial skin, but do not share the same range.
- Males have extensive red periorbital skin, unmarked dark brownish-grey upperparts, darker brown on the crown and nape, greyer throat and neck with pale feather fringes, while the underparts are dark grey, darkest on the under-tail coverts, with all of the feathers having darker centres and blackish shaft streaks. The flight feathers and tail are dark brown.
- Females are distinctive, with mottled upper- and underparts vermiculated with black, dark brown and buff, and some off-white U-to-V patterning on the belly and U-patterning on the lower neck.
- Juveniles are similar to adult females but have white barring below, black-and-white tips to the flank feathers, less extensive or no bare skin around eyes, and have a dull red bill and legs.

Distribution and habitat

- They are endemic to the south-east slopes of Mount Cameroon between 850 and 2 100 m above sea level.
- They are confined to primary and secondary montane forests provided there is dense undergrowth, along forest edges and in old fields adjacent to the forests.
- They avoid montane grasslands on the upper slopes of Mount Cameroon, but have been observed in savanna scrub habitats following forest burning.

Habits

They mostly occur in pairs, foraging among leaf litter. They are territorial and shy, will fly to trees to escape dogs, but otherwise run to the safety of thicket cover when alarmed. Their call is a short high-pitched 3-noted '*Kilu kilu kilu*' or slower '*Kee-ku kee-ku kee-ku*' whistle usually given from a concealed perch at dusk.

Food and feeding

Their diet consists of berries, grass seeds and insects.

Breeding

They breed during the dry season, laying their clutch from October to December. Nests are usually located in the savannas or grasslands adjacent to forest edges. One nest was lined with leaves and had a clutch of 4 plain creamy buff eggs. No further information on their breeding biology is available.

Status and conservation

They are reportedly locally common, especially on the southern slopes of the mountain down to 850 m above sea level. There is no evidence of marked population declines. Although much of the forests on Mount Cameroon have suffered damage, the species appears to be able to tolerate secondary forests. However, because of their limited range they are classified as Endangered (The IUCN Red List of Threatened Species v 2015-4). They may be adversely affected by deliberate burning, by man and/or volcanic activity, which causes the retreat of forest edges, and by excessive hunting.

Swierstra's Spurfowl
Francolin de Swierstra

Pternistis swierstrai

Francolinus swierstrai (Roberts, 1929), Mombolo, south-western Cuanza Sul district, Angola

ABOVE Male Swierstra's Spurfowl (photo by Dayne Braine)

LEFT Female Swierstra's Spurfowl (photo by Dayne Braine)

Classification

The Swierstra's Spurfowl is monotypic with no subspecies recognised.

Description

- They are a large, slightly sexually dimorphic brown-backed montane spurfowl with a diagnostic broad black breast band, white throat and broad supercilium.
- The bill and legs of both sexes are red.
- The male's black breast contrasts with the light throat while the white belly has broad buff central streaks with blackish margins.
- Females have a paler more rusty back, mantle and upper wings.
- Their underparts are buff with irregular black or brown blotches or bars which are concentrated on the upper belly to form a mottled band and are sparse on the lower belly.
- Juveniles are similar to the adult female but their throat and supecilium are pale buff, and the upperparts are streaked and barred with rufous-buff.
- Downy chicks are underscribed.

Distribution and habitat

- This uncommon Angolan endemic is resident in the highlands of western Angola, in the Bailundu highlands and Mombolo Plateau, and with isolated populations on the Chela escarpment and Tundavala in the Huila district and at Cariango in Cuanza Sul district.
- They are restricted to the undergrowth and forest edges of a few relict patches of evergreen montane forest, such as on Mount Moco and Mount Soque.
- They also occur on the grass- and bracken-covered rocky slopes of mountain sides and in tall grass savannas on mountain tops and gullies.

Habits

They are secretive and usually remain secluded in dense undergrowth, shrubbery and large ferns. When disturbed, they run deeper into cover or flushed individuals will fly into a tree. The call is shrill and harsh, increasing in volume then fading away, similar to that of Jackson's Spurfowl *P. jacksoni*.

Food and feeding

They forage among fallen leaves in the undergrowth. The diet includes the seeds of legume and grass species, as well as insects.

Breeding

Very little is known about the breeding habits of this species. They are suspected to breed during May to July.

Status and conservation

Their habitat has been reduced by forest destruction, due to the extraction for timber and by clearing and burning for subsistence agriculture, and as a consequence their populations are small. Hunting may be a further problematic threat. They are Vulnerable (The IUCN Red List of Threatened Species v 2015-4) and listed on CITES Appendix II.

Ahanta Spurfowl
Francolin d'Ahanta

Pternistis ahantensis

Francolinus ahantensis (Temminck, 1854), Ahanta, Gold Coast (Ghana)

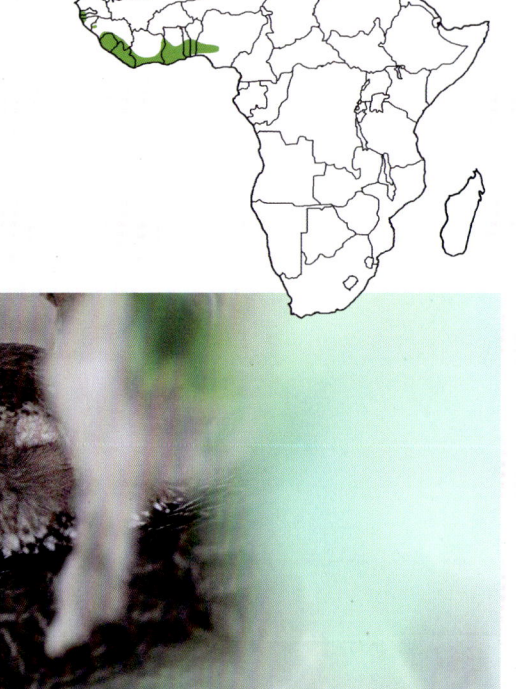

Ahanta Spurfowl in dense cover in the Bunso Arboretum, Ghana (photo by David Hoddinott)

Male Ahanta Spurfowl from Abuko Forest Reserve, The Gambia (photo by Bjarne Skov)

Classification

They are most closely related to Scaly Spurfowl *P. squamatus* and Grey-striped Spurfowl *P. griseostriatus*. Two subspecies are recognised:

P. a. hopkinsoni (Bannerman, 1934) in western Senegal, The Gambia and Guinea-Bissau

P. a. ahantensis (Temminck, 1854) in southern Guinea to south-western Nigeria

Description

- The Ahanta Spurfowl replaces the Scaly Spurfowl in West Africa, and is the only orange-billed spurfowl in its range.
- It is a large dark brown spurfowl, with an orange-red bill and orange legs and feet.
- They have a white throat and white edges to the feathers of their hind neck, back and underparts, appearing streaked at close range.
- The belly feathers are richly coloured and streaked with dark brown chestnut-edged buff.
- Juveniles are similar to adults but have arrow-shaped black streaks above.
- Downy chicks have a dark rufous-brown crown, nape and broad stripe down the middle of the back.
- The sides of the head are buffy-rufous with a rufous-brown eye-stripe extending back from the eye.

Distribution and habitat

- They are an endemic of West Africa with three separate populations: coastal areas of southern Senegambia and northern Guinea-Bissau; southern Guinea, Sierra Leone and western Liberia; north-eastern Ivory Coast and Ghana through the central Togo and central Benin to south-western Nigeria.
- Their preferred habitats are dense secondary vegetation at forest edges and in gallery forests, old plantations and cultivated patches.

Habits

They occur in pairs and small groups, and are generally shy and secretive. If flushed, they will typically fly to perch in a tree. They are most active and vocal at dawn, but will call at any time of the day and during moonlit nights. Their call is a loud raucous, squealing '*kee-kee-keree*', which is repeated several times and is higher pitched than that of the Scaly Spurfowl.

Food and feeding

Their diet consists mostly of seeds, small beans, cassava and large fruits, as well as insects, including termites.

Breeding

They breed from September until January in Senegambia and in late December in Ghana. Their nest is a scrape in the ground under thick cover and lined with leaves. The clutch of 4–6 eggs, sometimes as many as 12, are cream to pinkish-buff and measure 42.0 x 33.0 mm. Little is known about their incubation and chick-rearing.

Status and conservation

They are most common in north-eastern Ivory Coast and Ghana through central Togo and central Benin to south-western Nigeria, and rare and localised in the coastal areas of southern Senegambia and northern Guinea-Bissau.

Scaly Spurfowl
Francolin écaillé

Pternistis squamatus

Francolinus squamatus (Cassin, 1857), Cape Lopez, Gabon

Female Scaly Spurfowl in Kenya (photo by Jacques Pitteloud)

A pair of Scaly Spurfowls in a roadside forest clearing near Bishangari Lodge, southern Ethiopia (photo by Peter W. Hills)

A pair of Scaly Spurfowls along a road (photo by Aidan Kelly)

Scaly Spurfowl in typical forest edge habitat (photo by Nik Borrow)

Classification

They are most closely related to Grey-striped Spurfowl *P. griseostriatus*. Two subspecies are recognised:

P. s. squamatus (Cassin, 1857) in south-eastern Nigeria to the Democratic Republic of the Congo

P. s. schuetti (Cabanis, 1880) in eastern DRC to Uganda, Ethiopia, Kenya, Tanzania and Malawi

P. s. maranensis (Mearns, 1910), *P. s. usambarae* (Conover, 1928), *P. s. uzungwensis* (Bangs & Loveridge, 1931) and *P. s. doni* (Benson, 1939) are synonymised with *P. s. schuetti*

Description

- They are a dull, olive-brown spurfowl, characterised by pale buff-brown feather edges giving a scaly effect, particularly on the lower neck, orange-red legs and a red bill.
- The belly is plain brown with a scaly pattern and ill-defined dark shaft streaks margined buff.
- Juveniles are similar to adult females, but are more rufous with black arrow-shaped markings on the upperparts.
- Downy chicks have a dark rufous-brown crown, hind neck and broad mid-dorsal stripe.
- Their face and sides of the head are buffy white, with a single dark brown eye stripe.

Distribution and habitat

- Their distribution is across a belt of equatorial Africa between 10°N and 10°S, and between 800 and 3 000 m above sea level; from south-central Nigeria down to Gabon and far western DRC, across to Uganda, south-western Ethiopia and central Kenya.
- They prefer evergreen forests with dense undergrowth, forest clearings and the dense understory of bamboo and old plantations.

Habits

They are usually encountered in pairs or small groups which are secretive and skulking. When disturbed, they prefer to squat and then run to cover. They are active and vocal from before dawn and again at dusk, often into the evening and on moonlit nights. They call from a prominent perch, such as a termite mound. The call is a high-pitched, grating '*ke-rraak*' repeated 4–12 times, and increasing in volume to a crescendo.

Food and feeding

Their diet consists mostly of vegetable material including seeds and fruits, as well as snails, millipedes, termites, ants and other small insects. They will also eat cultivated crops, including cassava, sweet potato, peanuts and rice.

Breeding

They do not have a peak breeding season, but mostly lay from October to March, except for Gabon where they breed during June to August. Their nest is a shallow scrape under a tuft of grass or bush lined with grass and feathers. The clutch of 3–8, usually 6, oval hard-shelled pinkish-buff pitted white eggs measure 41.9 x 33.6 mm.

Status and conservation

Although locally common in places, even sometime more common than Double-spurred Spurfowl *P. bicalcaratus*, forest destruction and trapping may lead to local extinction.

Grey-striped Spurfowl
Francolin à bandes grises

Pternistis griseostriatus

Francolinus griseostriatus (Ogilvie-Grant, 1890), Quanza River, northern Angola

LEFT, ABOVE & BELOW LEFT Grey-striped Spurfowls in dense forest thickets (photos by Sean Braine)

RIGHT A roosting Grey-striped Spurfowl (photo by Tony Dowd)

Classification

They are most closely related to Scaly Spurfowl *P. squamatus* and Ahanta Spurfowl *P. ahantensis*. They are a monotypic species with no subspecies recognised.

Description

- The Grey-striped Spurfowl has rufous-chestnut-streaked plumage, a pale throat, unmarked face, red-orange bill and legs.
- The lower neck feathers and wing coverts are chestnut broadly vermiculated as in Scaly Spurfowl and Ahanta Spurfowl, but the belly is plain and the upper belly and flank feathers are chestnut and edged with greyish or creamy buff.
- They are paler than Scaly Spurfowl, which they replace in north-eastern Angola.
- They are differentiated from Red-necked Spurfowl *P. afer* by lacking bold black-and-white underparts and a bare red face and throat.
- Juveniles are rich cinnamon, with each feather having a blackish, not chestnut, centre and with black triangular-shaped markings above and below.
- Downy chicks are undescribed.

Distribution and habitat

- They are endemic to the escarpment zone and northern coastal plain of western Angola.
- There are two separate populations: southern Cuanza and western Milanje (Ndalo Tando, Punga Andongo, Dondo, Cuanza Basin and Cuanza Gorge); and another on the escarpment in the southern Benguela District and far north-eastern Huila District (Chingoroi and Caxito).
- They inhabit the dense thicket undergrowth of gallery and secondary forests and in riverine forests between 800 and 1 200 m above sea level.

Habits

They venture from dense cover to forage in grasslands and old cultivation lands, mostly former cotton fields, during the morning and late afternoon. When disturbed, they flush back to the forest edge. They make a duetting '*shwiii ke-ke-ke-ke*' advertisement call, as well as a softer '*fifififif*', repeated rapidly towards a crescendo. Their flight and alarm calls are a raspy '*kerak kerak*'.

Food and feeding

Their diet consists mostly of seeds and green shoots, as well as insects and other small arthropods.

Breeding

Their breeding habits are unknown.

Status and conservation

Little is known about their population status, and although they can be locally common, they are at least potentially impacted by forest destruction, including the clearance of undergrowth.

Double-spurred Spurfowl
Francolin à double éperon
Pternistis bicalcaratus

Tetrao bicalcaratus (Linnaeus, 1766), Senegal

Double-spurred Spurfowls from Nigeria (photos by Tasso Leventis)

Classification

They are most closely related to Clapperton's Spurfowl *P. clappertoni*, Heuglin's Spurfowl *P. icterorhynchus* and Harwood's Spurfowl *P. harwoodi*. Three subspecies are recognised:

P. b. bicalcaratus (Linnaeus, 1766) (including *P. b. ayesha* (Hartert, 1917) in western Morocco) and in Senegal to south-western Chad

P. b. adamaue Neumann, 1915 in southern Nigeria

P. b. ogilvie-granti (Bannerman, 1922) in southern Cameroon

Description

- The Double-spurred Spurfowl has an olive-green bill and legs, and lacks any bare facial skin.
- They have a chestnut nape, black frons and eye-stripe, a conspicuous white supercilium, streaked back and chestnut black-and-white tear-drop-marked breast and underparts.
- Females have smaller spurs.
- Juveniles are generally duller and the underparts are buff blotched with black.
- Downy chicks have a rufous brown crown with dark brown crown and eye stripes.
- They have a dark rufous-brown stripe down the mid-back, flanked with a buff stripe on either side.

Distribution and habitat

- They are endemic and common in drier areas of West Africa from Senegal to northern Cameroon and southern Chad.
- There are also two isolated populations in central and northern Morocco.
- They are the commonest spurfowl in West Africa outside of forests.
- Their preferred habitats are open grasslands, lightly wooded savanna and woodlands, palm groves, cultivation clearings including oil-palm, cassava and cocoa plantations, and croplands including maize, rice, cereals, groundnuts and cotton.
- In Morocco, they inhabit moist coastal open areas, groves and cultivation lands between forests and woodlands.

Habits

They are usually found in groups of 2–12 individuals, or larger flocks of up to 40 individuals. They feed in relatively open areas during the day, often seen on road tracks, particularly during the rainy season. They will seek cover in thickets and forest edges where savannas are burnt, and when disturbed they run to cover. They often call from a prominent perch such as a termite mound. The advertisement call is a series of loud raucous '*ke-rraak, keke-rraak*' repeated crowing, mostly at dawn and dusk.

Food and feeding

Their diet consists largely of fruits, roots, green leaves, seeds and agricultural millet, maize, rice and groundnuts. Animal food items include termites, caterpillars, ants, beetles, grasshoppers, small molluscs and frogs.

Breeding

They are monogamous and their breeding season is variable geographically, but mostly related to water availability and the rainy season. In Morocco breeding peaks in May and June. The nest is a shallow scrape sometimes lined with grass and feathers in patchy cover. The clutch of 5–7 smooth oval sandy or yellowish-buff eggs, sometimes with darker spots, measure 42.0 x 33.0 mm.

Status and conservation

Although common, they are susceptible to hunting pressures, habitat destruction and frequent burning.

Heuglin's Spurfowl
Francolin à bec jaune
Pternistis icterorhynchus

Francolinus icterorhynchus (Heuglin, 1863), Bongo, Bahr-el-Ghazal, Sudan

Female Heuglin's Spurfowl (photo by Nik Borrow)

Hueglin's Spurfowls in recently burnt open woodland in the Central African Republic (photo by Andy Pugh)

Male Heuglin's Spurfowl (photo by Cliff Dorse)

Classification

Heuglin's Spurfowl is most closely related to Double-spurred Spurfowl *P. bicalcaratus*, followed by Clapperton's Spurfowl *P. clappertoni* and Harwood's Spurfowl *P. harwoodi*. They are monotypic with no subspecies recognised.

Description

- They are easily distinguished by their yellow-orange bill and legs, and a yellow-brown bare facial skin patch behind the eye.
- They have a chestnut crown, brown back and pale cream underparts, which are heavily marked with black chevrons.
- Females are similar to males, but are slightly smaller.
- Juveniles are more heavily barred above.
- Downy chicks have a rufous-brown crown with a dark border.
- The sides of the head are buffy with a dark brown eye-stripe.
- Their back has a broad dark brown down the middle, with a buff stripe either side.

Distribution and habitat

- They are localised to the Central African Republic, northern Democratic Republic of the Congo, South Sudan and western Uganda.
- They are common in grasslands, open woodlands and adjacent agricultural fields.

Habits

They are usually encountered as pairs or small groups of up to 5 individuals. They can be difficult to see and when approached will run into cover or flush with frantic wing-beats into a tree. Calling males will often perch on a prominent site such as a tree branch or termite mound, and may be joined by the female. Adjacent territory males may answer the advertisement call. Their call is a harsh, slow '*kerak-kerak-kek*' or faster '*kerak-kerak-kerak-kerak-kerr*', similar to that of Double-spurred and Clapperton's Spurfowls, and mostly given just after sunrise and in the late afternoon.

Food and feeding

Their diet consists mainly of seeds and berries, as well as beetles, bugs, millipedes, termites and ants.

Breeding

They are monogamous and have an extended breeding season from April to November. The nest is a scrape under a bush or in thick vegetation. The clutch of 6–8 pale greyish buff eggs measure 42.0 x 33.3 mm. Little is known about their incubation or chick rearing.

Status and conservation

Similar to the related spurfowls of the vermiculated group, they are common to abundant in most of their range, but are probably impacted by hunting, habitat destruction and frequent habitat burning.

Clapperton's Spurfowl
Francolin de Clapperton

Pternistis clappertoni

Francolinus clappertoni (Children & Vigors, 1826), Bornu, (Chad / Nigeria)

Clapperton's Spurfowl calling from an elevated mound (photo by Tasso Leventis)

Perched Clapperton's Spurfowls (Left photo by Tasso Leventis, right photo by Nik Borrow)

Classification

They are most closely related to Harwood's Spurfowl *P. harwoodi*, followed by Double-spurred Spurfowl *P. bicalcaratus* and Heuglin's Spurfowl *P. icterorhynchus*. Six subspecies are recognised:

P. c. clappertoni (Children & Vigors, 1826) in Mali and western Sudan

P. c. koenigseggi (Madarász, 1914) in eastern Sudan

P. c. heuglini (Neumann, 1907) in south-western and south-central Sudan

P. c. sharpii (Ogilvie-Grant, 1892) in northern and central Ethiopia and Eritrea

P. c. nigrosquamatus (Neumann, 1902) in eastern Sudan and western Ethiopia

P. c. gedgii (Ogilvie-Grant, 1891) in south-eastern Sudan to north-eastern Uganda

In future, only two subspecies may be recognised:

P. c. clappertoni, including *P. c. heuglini* and *P. c. gedgii*; *P. c. sharpii*, including *P. c. koenigseggi* and *P. c. nigrosquamatus*

Description

- Clapperton's Spurfowl are easily distinguished from Double-spurred Spurfowl, Heuglin's Spurfowl and Chestnut-naped Spurfowl *P. castaneicollis* by their red facial skin and red legs.
- They have a fairly extensive white throat and neck, and white edges to the feathers giving a scaled appearance to the back and underparts.
- They differ from Harwood's Spurfowl by having a white supercilium, mostly black bill with a red base, and a variable black moustachial stripe.
- Juveniles are duller overall and are less distinctly marked.
- The downy chick is undescribed.

Distribution and habitat

- Their distribution is widespread across north-central Africa from far eastern Mali, central Niger, far north-eastern Nigeria, Chad, southern Sudan and northern South Sudan, north-eastern Uganda and western Ethiopia.
- They are generally patchily distributed but can be locally common to abundant.
- Their preferred habitats are within sandy grasslands, semi-arid bushy savanna, and often in adjacent fields and near water.

Habits

They are mostly seen in pairs and small groups, which are most active during the late afternoon. Males call from a conspicuous site such as a tree branch or termite mound. The call is a loud raucous '*kerraaak*' repeated crowing, with variations, at dawn and dusk, similar to that of Double-spurred and Heuglin's Spurfowls.

Food and feeding

Their diet consists of seeds, berries, insects and small molluscs.

Breeding

They are monogamous and their breeding season peaks from July to September. The nest is a well concealed scrape. The clutch of up to 9 thick-shelled yellowish-brown eggs have distinct pores and measure 43.0 x 33.0 mm. Little is known about their clutch size, incubation or chick rearing.

Status and conservation

They are locally common to abundant and probably are vulnerable to hunting, habitat destruction and frequent habitat burning.

Hildebrandt's Spurfowl
Francolin de Hilderbrandt

Pternistis hildebranti

Francolinus hildebranti (Cabanis, 1878), Voi, Teita district, Kenya Colony

Juvenile Hilderbrandt's Spurfowl (photo by Ian Little)

Male Hilderbrandt's Spurfowl on an elevated fence perch in Kenya (photo by Jacques Pitteloud)

Male Hilderbrandt's Spurfowl in scrub habitat in Tanzania (photo by Tasso Leventis)

Classification

They are most closely related to Natal Spurfowl *P. natalensis*. Three subspecies are recognised:

P. h. altumi (Fischer, GA & Reichenow, 1884) in western Kenya

P. h. hildebrandti (Cabanis, 1878) in eastern and central Kenya, northern and western Tanzania, south-eastern Democratic Republic of the Congo and north-eastern Zambia

P. h. johnstoni (Shelley, 1894) in south-eastern Tanzania, Malawi and northern Mozambique

P. h. altumi and *P. h. hildebrandti* might be synonymised, while *P. h. johnstoni* might be synonymised with *P. h. fischeri* Reichenow, 1887

Description

- They are a member of the vermiculated spurfowl group and their rufous-buff underparts are distinctive.
- Males are similar to Natal Spurfowl, but have paler underparts which are clearly marked with large black spots. However, their ranges also barely overlap.
- Males are also similar to Clapperton's Spurfowl *P. clappertoni*, but they lack the red facial skin and the obvious white supercilium.
- Females have distinctively plain rusty underparts.
- Both sexes have red legs.
- Juveniles are similar to adult females, but are more heavily barred blackish above and below.
- Downy chicks have a rufous-brown crown with dark stripes and a dark brown stripe from the front of the eye back to the nape which is more pronounced than in other spurfowl chicks.

Distribution and habitat

- They are endemic to East Africa where they are thinly distributed from Kenya through most of Tanzania, north-eastern Zambia and Malawi.
- They prefer extensive dense thickets and scrub on grassy hill slopes, acacia woodlands, as well as bracken-briar edges of evergreen forest and montane heath.

Habits

They usually occur in pairs or small groups and are wary, sitting particularly tight when approached. Their raucous crowing '*KEK kek kek kerak*' call, with the first note the loudest, is similar to that of Natal Spurfowl, and is more often heard than they are seen.

Food and feeding

Their diet consists of seeds, bulbs, tubers, as well as insects and their larvae.

Breeding

Although their breeding season can extend from March to November, peak breeding is in June and July. The nest is a small scrape, lined with grass and leaves, usually well concealed within vegetation. The clutch of 4–8 creamy white to pale brown eggs measure 39.8 x 31.9 mm.

Status and conservation

Because of their localised, uncommon occurrence, little is known about their status and conservation.

Natal Spurfowl
Francolin du Natal

Pternistis natalensis

Francolinus natalensis (Smith, A, 1834), Durban, South Africa

Courting male Natal Spurfowl in the Kruger National Park, South Africa (photo by Jessie Walton)

Female Natal Spurfowl with chicks (photo by Ian White)

GAMEBIRDS

Calling Natal Spurfowl (photo by Ian White)

Perched Natal Spurfowl (photo by Ian White)   Juvenile Natal Spurfowls (photo by Ian White)

Female Natal Spurfowl (photo by Andre Botha)

Conspicuous calling territorial male Natal Spurfowl (photo by Andre Botha)

Classification

They are most closely related to Hildebrandt's Spurfowl *P. hildebrandti*. There are two subspecies recognised:

P. n. neavei (Mackworth-Praed, 1920) in southern Zambia and western Mozambique

P. n. natalensis (Smith, A, 1833) in Zimbabwe and north-eastern South Africa

They are reported to hybridise in the wild with Swainson's Spurfowl *P. swainsonii* and Red-necked Spurfowl *P. afer*.

Description

- The size difference between the similar sexes is marked, and larger males have stout leg spurs.
- Their overall plumage is dark brown, with white and dark brown vermiculations across the breast and flanks.
- The crown and sides of the head are brown, grading downwards into black and white.
- The restricted white on the throat is less conspicuous than that of Cape Spurfowl *P. capensis*.
- They lack bare skin on the face and throat.
- The legs are bright orange, particularly during the breeding season.
- The bill is also orange with a dull greenish base.
- Juveniles resemble adults but are paler, with greyish-brown upperparts with dense vermiculations and indistinct rufous-buff barring.
- The underparts are buffy with faint black barring, with each feather having a conspicuous white shaft streak.
- Downy chicks are rufous-brown above, with prominent darker streaks on the crown and back.
- The sides of the head are buffy, with a dark brown eye stripe extending from the back of the eye.
- The underparts are buffy, with light rufous-brown on the breast.

Distribution and habitat

- They occur across north-eastern South Africa, Swaziland, inland Mozambique, Zimbabwe, eastern Botswana and south-eastern Zambia. Although most common in the lowveld and drier savannas, they also venture into riverine bush among escarpment grasslands.

- They inhabit savannas and woodlands with scrub thickets, riverine vegetation and montane forest edges to coastal forest.
- They have expanded their range along watercourses invaded by the alien Black Wattle *Acacia mearnsii*.
- They also use modified habitats, including cultivated land edges adjacent to good cover, and firebreaks cut through commercial plantations.

Habits

Although sometimes in groups of up to 10 birds, they are usually encountered as scattered individuals because coveys are loose and individuals disperse when disturbed. They will congregate where food is abundant, such as in livestock feedlots where waste grain is available. When disturbed, they normally run into thicker cover, or flush as individuals rather than as a covey, flying a short distance before falling back into cover. A flushed bird will often land on a perch and freeze, hoping to avoid detection. Like most other spurfowls, they often fly to their overnight roosting site, thus avoiding being tracked by predators. Where cohabiting with humans, such as at conservation area campsites, they will become tame, keenly accepting or even soliciting food items. Their advertisement call '*kak-keek, kak-keek, kak keeeee*k' at dawn and dusk is similar to that of Cape Spurfowl and is often the only evidence of their presence in the area. They also have a high-pitched, mewing contact call when individuals are separated.

Food and feeding

They usually feed during the early morning and late afternoon, and tend to forage under the cover of vegetation where they scratch in leaf litter and loose soil to detect food items. Bulbs, roots, seeds, berries and fruits are their main diet, but they also eat fallen grains such as sorghum, cowpeas and maize. During the austral summer they also feed on beetles, termites, grasshoppers and caterpillars. They also scratch through rhinoceros (Rhinocerotidae) and African elephant *Loxodonata africana* dung for undigested seeds, insects and their larvae.

Breeding

The breeding season is extended, particularly in the northern parts of their range, where breeding has been recorded during all months of the year. Peak breeding is from December to February in South Africa and March to May in Zimbabwe. The nest is a shallow scrape lined with roots and grass and is concealed beneath dense entanglements of scrub. The clutch of 5–7 rounded cream or buff-coloured eggs measure 41.9 x 34.2 mm. Incubation by the female lasts 20–22 days. Chicks can flutter-fly at 10–14 days.

Status and conservation

There are no reasons for concern regarding the conservation of this species. Potential threats are habitat modification and fragmentation by woodland removal and frequent burning, which has led to declines in their numbers in parts of South Africa. They are sustainably hunted in many parts of their range, along with Helmeted Guinefowl *Numida meleagris* and Swainson's Spurfowl.

Hartlaub's Spurfowl
Francolin de Hartlaub

Pternistis hartlaubi

Francolinus hartlaubi (Barboza du Bocage, 1869), Huila, southern Angola

Male Hartlaub's Spurfowl (photo by Cliff Dorse)

Pair of Hartlaub's Spurfowl, female on the right (photo by Dayne Braine)

Male Hartlaub's Spurfowl (photo by Dayne Braine)

Male Hartlaub's Spurfowl (photo by Maans Booysen)

Female Hartlaub's Spurfowl (photo by Cliff Dorse)

Classification

Hartlaub's Spurfowl is most closely related, although relatively distantly, to Djibouti Spurfowl *P. ochropectus* of East Africa. Previous suggestions for any subspecies are not supported, thus there are no subspecies recognised.

Description

- This sexually dimorphic spurfowl has a relatively small head and an unusually large down-curved beak.
- The beak and legs are dull yellowish-brown.
- Males have a dark greyish-brown crown, a conspicuous white-over-black eye-stripe and chestnut ear coverts.
- They have contrasting dark brown and faintly barred back plumage, with pale belly feathers, heavily streaked with brown.
- The lower surface of the male's tail feathers is blackish brown, distinctly barred black and white, which is conspicuous in flight and during courtship display.
- Females have a grey-brown head, and orange-brown eye-stripe, cheeks, chin and belly, with dark grey-brown, boldly marked back plumage.
- Neither of the sexes has leg spurs, only small rounded bumps.
- Juveniles have a buffy forehead and a brown crown tinged with rufous.
- The side of the head is buffy with a faint greyish stripe behind the eye.
- The upperparts are buffy-grey, vermiculated with black and white, with each feather having a white shaft streak.
- The breast is barred black and white, grading to buffy-grey on the belly, with each feather having a white shaft streak.
- The plumage of the chick has not been described.

Distribution and habitat

- This near-southern African endemic is scattered in central and northern Namibia, particularly on the Namibian escarpment, and slightly into south-western Angola.
- They are confined largely to granite and sandstone outcrops surrounded by semi-desert steppe.
- The vegetation within and immediately adjacent to these outcrops is invariably dense, mixed grass-scrub, with heavy undergrowth, on sandy soils.

Habits

They are sedentary and territorial. They are not easily flushed, preferring to seek refuge in gaps between boulders. If flushed, they fly quickly, often uttering a rapid, 3–5 second chattering '*krak*'. Pairs are difficult to observe as they dart between rocks and among vegetation. They remain monogamous throughout the year, and the pair-bond assists with conflicts with conspecifics which are resolved by call contests and postural displays rather than by physical confrontation. Each partner helps to demarcate the territory with a complex, raucous, antiphonal (duet-like) advertisement call initiated by the female. They have a complex and elaborate courting ritual, thereafter the bond is maintained by interactive vocal contact. Their ventriloquial daily advertisement calls are uttered at sunrise and comprise a closely synchronised, antiphonal duet by the male and female alternating rapidly, '*wiich ter wideo, wiich ter wideo, wiich ter wideo …*'.

Food and feeding

The pair ventures from the security of boulder outcrops to feed in openings adjacent to their refuge. The powerful bill is used to scrape out corms and bulbs, and large areas of soil can be upturned while foraging. They eat mainly corms, bulbs and rhizomes of small sedges, particularly 'uintjies' *Cyperus edulis*, as well as seeds, fruits, beetles, grubs, larvae and land molluscs.

Breeding

Most breeding records are from May to August. The nest scrape is well hidden between boulders. Clutch size is relatively small, with only 2–4, usually 3, rounded plain cream eggs, measuring 41.6 x 30.2 mm. Incubation lasts for 23 days. Also unusual in gamebirds is that the chicks are cared for by the male. Chicks can flutter-fly at 3 days and fledge at 12 days. Sexual dimorphism begins to show after 100 days.

Status and conservation

Because of their low population densities and confinement to suitable habitat, the local population size and distribution might fluctuate according to droughts or due to overgrazing and are probably influenced by predators. Their habitats might be further threatened by granite-mining activities. However, there is no evidence that distribution patterns have changed to any large degree in recent times.

Harwood's Spurfowl
Francolin de Harwood

Pternistis harwoodi

Francolinus harwoodi (Blundell & Lovat, 1899), Aheafeg (= Ahaia Fej & Haiafegg), Shoa, Ethiopia

Male Harwood's Spurfowl
(photo by Nik Borrow)

Male Harwood's Spurfowl (photo by Nik Borrow)

Male Harwood's Spurfowl (photo by Adam Riley)

Classification

They are most closely related to Clapperton's Spurfowl *P. clappertoni* followed by Double-spurred Spurfowl *P. bicalcaratus* and Heuglin's Spurfowl *P. icterorhynchus*. They are monotypic with no subspecies recognised.

Description

- The Harwood's Spurfowl is a grey-brown spurfowl with bare red skin around the eyes, a red bill and red legs.
- They are darker than Clapperton's Spurfowl, with less red skin around the eyes and largely red, not mostly black, bill.
- The breast is vermiculated black and pale buff.
- They are also differentiated from Erckel's Spurfowl *P. erckelii* by their reddish, not black, bill, red, not yellow, legs and vermiculated, not streaked, underparts.
- Females are similar to males but are slightly paler and browner below with a less streaked belly.
- Juveniles are similar to adults and the downy chick is undescribed.

Distribution and habitat

- They are a very range-restricted species in the gorges of the Jemmu valley, Blue Nile River and its tributaries in the highlands of central Ethiopia.
- They prefer dense and extensive *Typha* reedbeds along streams, semi-arid scrub and thickets along rivers and on scrubby hill slopes, foraging on adjacent fields, often near the reeds.

Habits

They are not overly shy, but when disturbed, they will fly back into reedbeds, which they also use for shade during the heat of the day. They roost in trees within the reedbed or in the reedbed itself. Their call is a harsh raucous '*ko reek*' crowing, usually given in sets of 3 well-spaced calls at dawn and dusk, similar to that of Clapperton's Spurfowl.

Food and feeding

They will venture out into open agricultural lands, such as sorghum, to forage. Their diet consists of tubers, berries, grass and sorghum seeds, as well as worms and termites. Although they consume agricultural seeds, they are not considered to be a pest.

Breeding

Very little is known about their breeding biology, with the breeding season possibly being from November to February, and the clutch size being between 3 and 10 eggs.

Status and conservation

Although they are classified as Vulnerable (The IUCN Red List of Threatened Species v 2015-4), they are locally common at several sites. Habitat destruction by clearing for agriculture and timber harvesting for fuel and construction, and the injudicious use of fire, may affect their local range. They are also trapped using leg snares, as they are reported to be good food, and also for their reputed medicinal properties.

Red-billed Spurfowl
Francolin à bec rouge

Pternistis adspersus

Francolinus adspersus (Waterhouse, 1838),
Upper Kuiseb River, Namibia

ABOVE Group of Red-billed Spurfowls (photo by Michael Poliza)

FAR LEFT Male Red-billed Spurfowl (photo by Eric Stockenstroom)

LEFT Male Red-billed Spurfowl (photo by Yathin S Krishnappa)

SPURFOWLS

Calling male Red-billed Spurfowl
(photo by Ian White)

Female Red-billed Spurfowl
(photo by Ian White)

Red-billed Spurfowl portrait
(photo by Ian White)

Classification

They are most closely related to Natal Spurfowl *P. natalensis* and Cape Spurfowl *P. capensis*. Two subspecies are recognised:

P. a. adspersus (Waterhouse, 1838) in south-western Angola, north-western Namibia and northern Botswana

P. a. mesicus (Clancey, 1996) in two isolated populations in mesic areas of the Waterberg plateau, north-central Namibia and in the eastern Caprivi, western Zimbabwe and south-western Zambia.

Description

- Although in the vermiculated group of spurfowls, their feathers are not flecked, but have pale white bars.
- They are medium sized, with conspicuous and distinctive yellow eye rings and finely barred underparts.
- The bill and legs are orange-red, and adult males have formidable spurs.
- Females are smaller than males, with smaller blunt spurs.
- Juveniles are lighter brown, with buff bars and streaks, and do not have the yellow skin around the eye.
- Downy chicks have a rufous-brown crown with a darker stripe.
- The sides of the head are buffy, with a dark stripe extending backwards from the eye.
- A broad dark stripe extends down the mid-back, flanked by narrower buff stripes.
- The rest of the underparts are pale yellowish-buff, darker on the breast.

Distribution and habitat

- They are common in the drier, sandy Kalahari and Zambezian areas of south-western Africa, extending marginally into southern Angola and south-western Zambia.
- They are abundant throughout central and northern Namibia, the western corner of Zimbabwe, and throughout Botswana, except for the drier south-western Kalahari region and far eastern Botswana.
- Although they generally inhabit relatively dry savannas, they are also abundant on floodplains in the Okavango Delta and in the fossil-like beds of the Makgadikgadi Pan.
- Typical savanna habitats are *Baikaiea*, *Senegalia* and mixed woodlands, low scrub, thickets interspersed with open ground, and edges of woodlands on Kalahari sand, usually not far from water, and frequently along water courses.
- There has been a southward extension of their distribution, during the late 1900s, into southern Botswana and the Northern Cape in South Africa.

Habits

They are usually seen in pairs or groups of up to 12 birds. Although often shy and elusive, they can be bold at sites in association with humans. They are known to venture into smaller rural villages, where they free-range feed with domestic poultry. When disturbed, they run rapidly into nearby cover or flush with a harsh alarm call, often seeking a perch to avoid intruders. They are most active at dawn and dusk, resting under the shade of broad-leafed scrub during the heat of the day. Often, their scratching for food items in leaf litter under scrubby cover will disclose their presence. Their '*ka-wak-wak-wak, ka-krr-krr-krr*' call, heard mostly at dusk and dawn, and often well before daylight, is similar to that of Cape Spurfowl and Natal Spurfowl, but slightly more strident.

Food and feeding

Their diet consists largely of bulbs and corms in the dry season, but they will also feed on fruits, berries and seeds. Their close association with relatively lush areas that are adjacent to water mean that they disperse or concentrate depending on seasonal rainfall. They often associate with large game, particularly at waterholes, where they scratch through droppings in search of undigested seeds and associated invertebrates.

Breeding

They are a solitary nester and are probably monogamous. Although they breed throughout the year, usually after substantial rainfall, peak breeding activity is from April to September in Namibia, Botswana, Zambia and Zimbabwe, and during November to January in the Northern Cape. The nest is a scrape in the ground, sparsely lined with dry vegetation and placed among grass tufts or under a bush. It is, however, often relatively poorly concealed compared to that of most other spurfowl nests. The average clutch is 5 eggs, but can range from 4–10 eggs, which measure 42.2 x 33.4 mm. The eggs are dull yellowish, dirtying as the incubation progresses, and are particularly thick shelled and pitted, resembling small guinefowl eggs. The incubation period is 22 days. Juveniles attain adult plumage at about 3 months and the male's spur develops at about 5 months.

Status and conservation

Their tolerance of human activity and preference for somewhat modified habitats, coupled with their general abundance, place them beyond consideration for conservation action. The only threat to the otherwise secure populations is degradation or clearing of habitat by and for livestock. They are hunted as an opportunistic addition to game animals and the offtake levels should be carefully monitored to ensure that they can be sustained.

Cape Spurfowl
Francolin criard

Pternistis capensis

Tetrao capensis (Gmelin, JF, 1789), Cape of Good Hope, South Africa

ABOVE Cape Spurfowl shaking off a dust bath (photo by Jessie Walton)

LEFT Male Cape Spurfowl (photo by Davide Gaglio)

BELOW Cape Spurfowl in coastal fynbos habitat (photo by Davide Gaglio)

Female Cape Spurfowl (photo by Maans Booysen)

Preening Cape Spurfowl (photo by Ian White)

Female Cape Spurfowl with a juvenile (photo by Davide Gaglio)

Classification

The Cape Spurfowl is monotypic with no subspecies recognised.

Description

- They are dark overall, with white streaking and flecking, particularly on the breast and flank feathers.
- When alarmed, they raise their dark crown feathers.
- In flight, the tail appears blackish.
- The legs are dull orange-red and males have large spurs, often two on each leg, one above the other.
- The upper bill is dark brown with an orange-red base and the lower is dull orange.
- There is no bare skin on the face or throat.
- Juveniles are similar to adults, but are browner above and grey below, with less distinct markings.
- Downy chicks have a rufous-brown crown and nape, narrowly bordered with a darker stripe on each side.
- The sides of the head are buffy-white, with a blackish brown eye-stripe.
- The back is brown with a dark brown median stripe bordered by buffy stripes.
- The underparts are buffy, but darker on the breast.

Distribution and habitat

- They are endemic to southern African fynbos and karoo habitats, from the mouth of the Olifants River in the west to Alexandria near Port Elizabeth in the east. Isolated populations occur northwards into the Karoo, up the Western Cape coast and along the lower Orange River, from Upington to the Atlantic coast.
- They inhabit scrubby heath, especially coastal fynbos (standveld and renosterbos), and in sheltered shrubs along streams and rivers.
- They are also fond of stands of introduced Australian acacias.
- They rarely wander far from escape shelter and their densities are affected by the placement, size and connectedness of fragmented natural habitats and patches of alien bush.

Habits

They are conspicuous and occur in pairs or small groups, averaging 5 birds in the non-breeding season. When disturbed, they escape by running, flying only when pressed. Despite being alert to danger, where they are accustomed to human presence such as in campsites and parks, they can be bold, regularly approaching humans and human dwellings to scavenge for food. Overnight roosting sites are invariably in dense scrub patches, reed beds or patches of Palmiet *Prionium serratum* in wetland sponges, where they huddle together on the ground or, occasionally, perch in trees. Coveys fly across river estuaries at dawn and dusk, using the estuary as a barrier to terrestrial predators between their daily activity sites and their roost site. The advertisement call is a loud crowing '*kak-keek, kak-keek, kak-keeeeek*', with the second syllable accented, similar to Natal Spurfowl *P. natalensis*, and is usually uttered in the early morning and late afternoon. Juveniles give a single long tonal whistle contact call, decreasing in frequency.

Food and feeding

Their diet is primarily bulbs, corms, seeds, berries and fallen grain during the non-breeding season, with small snails, termites, ants and insects also eaten during the breeding season. They also feed on fallen fruits, such as grapes, apples and pears.

Breeding

Breeding occurs from August to January, peaking during September and October. Pairing is characterised by males vigorously chasing each other to determine dominance. The nest is a scrape lined with grass and is usually well concealed in a patch of fynbos or under a bush. The clutch is 4–8 eggs, but can be as many as 14, possibly when 2 females contribute to the clutch. The round to oval eggs are brownish-cream to pale pink and measure 47.0 x 37.0 mm. Incubation by the female lasts 22–23 days. Chicks can fly short distances at 12 days old.

Status and conservation

There is no evidence of recent population declines or range contraction. They benefit from habitat transformation by colonising alien vegetation patches and suburban parklands. However, in agricultural land, such as wheat, vineyard and deciduous fruit farms, they are dependent on roosting and nesting sites in connected patches of fragmented natural habitats.

PART 4

Quails

QUAILS

Classification

Old World quails occur widely throughout Africa, Europe, Asia and Australia and belong to the Tribe Coturnicini. The three African quail species belong to the family Phasianidae and to two genera:

Coturnix Garsault, 1764, containing two species:

- Common Quail *C. coturnix* of Africa, Europe and central Asia
- Harlequin Quail *C. delegorguei* of Africa and southern Arabia

Synoicus (Excalfactoria) (Linnaeus, 1766), containing one species:

- Blue Quail *S. adansonii* are widespread through sub-Saharan Africa and are a sister species of the King or Blue-breasted Quail *E. chinensis* of Asia, New Guinea and north-eastern Australia.

Description

- The back plumage of quails is characteristically rufous-chestnut, strongly marked with a white, longitudinal shaft streak and transverse white and black bars.
- Their tail has, depending on the species, 8–12 feathers (normally 14 in partridges, francolins and spurfowls, and 16 or more in pheasants and grouse).
- Their legs are yellow and lack spurs.
- Males of the three African species have musical, three-note advertisement calls, which are used presumably to notify each other of their territorial domains and to attract mates.

Natural history

The natural history of the three African quails is relatively poorly understood. The biology of even the most common species, the Common Quail *Coturnix coturnix*, in the wild is still not fully understood, making conservation and management of the species relatively challenging. In fact, much of the information on their biology is derived either from anecdotal observations or from observations of birds reared and kept in captivity.

These quails have complex movement and spatial occurrence patterns. Furthermore, the geographical ranges of all three species, unlike those of most other African gamebirds, extend outside of the borders of Africa.

PREVIOUS PAGE Male Harlequin Quail (photo by Hugh Chittenden)

Again, unlike Africa's other gamebirds, excluding sandgrouse, all of which venture no more than a few kilometres from the area where they were hatched, quails move or migrate within, and in some cases outside, of Africa. Some nomadic African populations of Common Quail move thousands of kilometres, from as far north as Kenya down to the Western Cape in South Africa.

Evolutionary placement

- Anatomically, quails are the smallest of the African gamebirds, yet their skeleton is remarkably similar to that of the Wild Turkey *Meleagris gallopavo*, which is the largest gamebird, being more than a hundred times the size of the Common Quail.
- Genetic studies suggest that the nearest relative of the three African quails is the Madagascan Partridge *Margaroperdix madagarensis*.
- The quails form a genetically quite distinct branch, well up in the evolutionary tree for galliform birds, along with the Madagascan Partridge, the African spurfowls *Pternistis* species, the Bush Quails *Perdicula* species of India, and the Sand and See See Partridges *Ammoperdix* species of northern Africa and the Middle East.
- They are very distantly related to the New World quails (Family Odontophoridae), which are genetically closer to the guineafowls (Numididae).
- Some other Asiatic partridges (for example, *Rollulus* spp. and *Arborophila* spp.) and the Udzungwa Forest Partridge *Xenoperdix udzungwensis* and Rubeho Forest Partridge *X. obscuratus* from Tanzania branch off near the base of the gamebird evolutionary tree.
- They are the relicts of the ancestral phasianine galliform from which the balance of pheasants, partridges and Eurasian quails are derived.
- They and other galliforms are not related to the buttonquails (Turnicidae), with which they co-occur in Africa, and which are, in turn, genetically closest to the shorebirds (Charadriiformes).

Conservation

Common Quail and Harlequin Quail are extensively hunted or netted for food by humans. Common Quail are also farmed in large numbers for table meat or to produce eggs. Where hunted, they are often artificially stocked on game farms or to supplement wild populations.

Common Quail
Caille des blés

Coturnix coturnix

Tetrao coturnix (Linnaus, 1758), Sweden

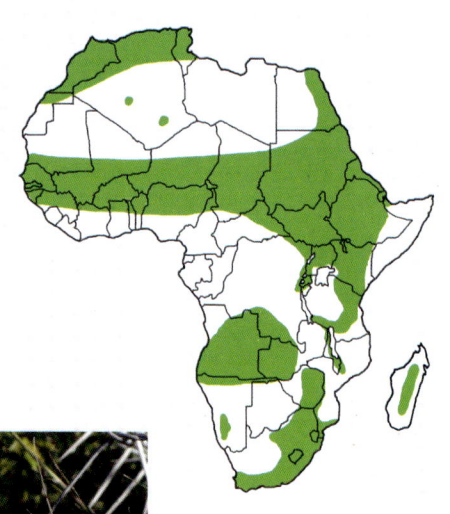

Common Quail (photo by Maans Booysen)

Common Quail in Dullstroom, eastern South Africa (photo by Ian Little)

Common Quail in Kenya (photo by Lorenzo Barelli)

Classification

Five subspecies are recognised:

C. c. coturnix (Linnaeus, 1758) in the British Isles and north-western Africa, east to east-central Russia and east-central India, possibly to Bangladesh

C. c. conturbans Hartert, 1917 and *C. c. inopinata* Hartert, 1917 on the Azores and Cape Verde Islands, respectively

C. c. africana Temminck & Schlegel, 1849 in sub-Saharan Africa, from Ethiopia, Democratic Republic of the Congo (DRC) and Uganda to southern Angola and South Africa, and also on Madagascar

C. c. erlangeri Zedlitz, 1912 in eastern Zimbabwe and adjacent Mozambique to Malawi, north-eastern Zambia (Nyika Plateau), Tanzania, eastern DRC, Uganda, Kenya, Sudan and Ethiopia

Description

- They are the plainest of the African quails.
- Their back plumage is rufous-brown, mottled with black.
- Each feather has a lance-shaped, buffy-white streak and two or more similarly coloured bars.
- The flanks are broadly streaked with black and buff, the breast is pale rufous-brown and the belly is dull grey.
- They have no bold facial or throat markings and there is relatively little difference in colour and plumage pattern between the sexes.
- Males have a narrow black, anchor-shaped throat patch compared with the white throat of females.
- Juveniles resemble adult females, but the flanks are less heavily streaked and the breast has brownish spots.
- Downy chicks are rufous-brown above, with two dark streaks on the crown which merge to form a single streak down the back. This streak is flanked by two additional dark streaks that merge with it on the rump.

Distribution and habitat

- Their distribution and movements are complex and variable.
- Northern populations are largely migratory.
- In south-eastern Africa, some populations are nomadic residents, whereas others migrate within the subcontinent, only rarely venturing into the Northern Hemisphere.

- The winter migration of the southern African population is into southern DRC, Angola, western and northern Namibia and western Zambia.
- It is uncertain if any proportion of the Northern Hemisphere population regularly visits southern Africa.
- Favoured habitats are grasslands, usually less than 50 cm tall, agricultural crops (particularly cereals), fallow weedy fields, and pastures (such as lucerne, clover, Eragrostis, Kikuyu and rye grass).
- Habitat attractiveness is determined by the abundance of insects in summer and seeds in winter, and by the type of cover afforded.
- Grasslands that are seldom burnt or grazed, or that are burnt too frequently and/or overgrazed, are not favoured.
- They also avoid the drier, hotter parts of Africa and areas with heavy bush and tree cover.

Habits

The proportion of the population which is migratory, nomadic or resident is uncertain. Changing local conditions may affect the faithfulness of birds to a particular area, and movement patterns may vary from place to place and from year to year. Their advertisement call is a ventriloquial '*whit wit-wit*', uttered repeatedly from within cover, which is slower and less evenly spaced than that of Harlequin Quail *C. delegorguei*. The flush call is a '*crwee-crwee-crwee*' trill. Unless flushed, they run swiftly through the grass in a hunched position when approached. Flocks of migrating birds usually fly on dark, moonless nights, and their passage can usually only be detected by the constant fluttering sounds of their wings.

Food and feeding

Their diet includes seeds (mostly of grasses, grains and weeds), flower buds, green leaves, and some smaller fruits, rhizomes and tubers (e.g. Cyperaceae). During the summer the diet is supplemented with invertebrates, particularly caterpillars and grasshoppers.

Breeding

C. c. erlangeri is a summer-breeding intra-African migrant resident in East Africa, while *C. c. coturnix* is a non-breeding Palearctic migrant. The nest is a scrape in the ground, lined with grass and rootlets, usually well concealed in dense herbaceous or grassy cover, sometimes in crops. The clutch size is 5–8 eggs, but can be as many as 12. The pointed oval eggs, which measure 30.3 x 23.4 mm, are smooth creamy-yellow with bold spotting and blotches of olive, dark brown and black. Incubation by the female lasts for 17–20 days.

Status and conservation

Because of the uncertainty of their movements, it is difficult to develop an effective conservation strategy for Common Quail. However, they have benefited from crop farming in some regions. They are prized as quarry by wing-shooters when they gather in high concentrations. Unfortunately, because they concentrate to breed during summer in eastern and southern Africa, hunting coincides with their breeding season. Therefore, hunting should be delayed to allow successful completion of breeding. Elsewhere in Africa, migrating flocks are heavily exploited for food by netting. Other landscape issues affecting the habitat availability for this species are expanding commercial afforestation, overgrazing and expanding rural human settlement.

Harlequin Quail
Caille arlequine
Coturnix delegorguei

Coturnix delegorguei Delegorgue, 1847, Oury, Upper Limpopo River, Limpopo Province, South Africa

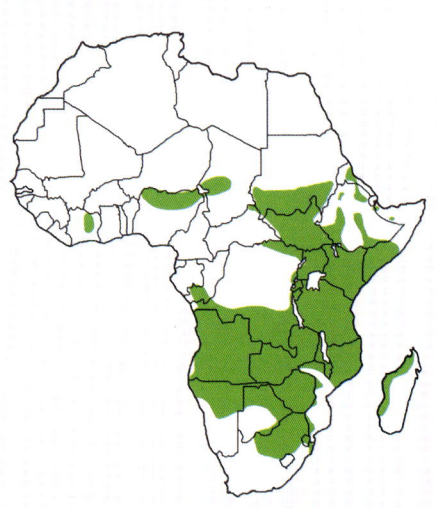

Female Harlequin Quail from São Tomé (photo by Tasso Leventis)

Calling male Harlequin Quail (photo by Richard Grant)

Classification

Three subspecies are recognised:

C. d. delegorguei Delegorgue, 1847 found from the Ivory Coast to Ethiopia (and is apparently the only migrant quail to West Africa) and south to South Africa and western Madagascar

C. d. histrionica Hartlaub, 1849 on São Tomé in the Gulf of Guinea

C. d. arabica Bannerman, 1929 in southern Arabia

Description

- They are smaller and more sexually dimorphic than Common Quail *C. coturnix*, but not as striking as in Blue Quail *Synoicus adansonii*.
- The white throat and thin white collar of the male are as conspicuous as those of male Blue Quail, but differ in having a white stripe above the eye.
- The black on the throat of males is anchor-shaped, with the shaft running down the centre of the throat and the hooks curling up behind each eye.
- Males are darker overall than male Common Quail and have chestnut and black underparts which are distinctive.
- Females are paler overall than males and are similar to female Common Quail, but lack any dark marking on the throat and their underparts are darker.
- Juveniles resemble adult females, but have a light brown crown with a pale streak and a light brown nape. Their underparts are pale grey, with the breast spotted and barred with dark brown, and each feather has a white shaft streak at its tip.
- Downy chicks have a brownish crown and nape with a buff streak. Their back is also brown, with a darker stripe down the mid-back that is flanked by narrower stripes of buff and black. The belly is buffy yellow.

Distribution and habitat

- They are widespread and locally common to abundant throughout south-central and eastern Africa, excluding forest belts and arid south-western Africa, and are scattered and less common in north-central and West Africa.
- They are both a nomadic resident and an intra-African migrant.
- They are uncommon migrants to western Madagascar.
- Rainfall plays a role in determining their movements. They can be irruptive in their presence locally during a summer, and then may fail to return there for several years.
- They prefer short (<30 cm), rank wooded and flooded grasslands, occasionally with scattered bush cover, and are not as common in highland grasslands as Common Quail.

Habits

They are gregarious, often found in coveys of 6–20 individuals during the non-breeding season. Large migrating flocks sometimes move on foot during both the day and night. High adult mortalities during migration occur where birds are killed or injured on roads by cars, or when birds fly into the windows of buildings. They are secretive, spending most of the day concealed within cover. They are reluctant to flush, preferring to sit tight until approached and then bursting into fast, low flight before dropping back into cover. Their three-note call is repeated monotonously, similar to Common Quail, but is less connected, a more deliberate '*wit, wit, wit*'. Calling males stand erect, displaying chestnut underparts and a distinct throat pattern. Most calling is during the morning and late afternoon, and after rain or when a storm is looming. The flush call is a squeaky '*kree-kree-kree*'.

Food and feeding

They usually forage within cover, but sometimes at the edges of roads or tracks. Their diet includes the seeds of grasses, weeds and some agricultural crops, e.g. Goose Grass *Eleusine* sp., bristle grasses *Setaria* spp. and Sorghum *Sorghum purpureosericum*. The breeding season diet is supplemented with grasshoppers, caterpillars, termites, ants, beetles and small molluscs.

Breeding

The breeding season is variable, linked to heavy rainfall. They may nest in loose colonies and the males may be polygynous. The nest is a small bowl with sparse lining, usually located in grasslands with well-spaced tufts or weedy fields. They lay 4–8 creamy eggs, and occasionally up to 20 eggs are probably laid by more than one hen, which are creamy, speckled, spotted and blotched with dark brown. They measure 29.2 x 22.4 mm. Incubation, by the female lasts 14–18 days.

Status and conservation

Although locally abundant in some years, the relative absence of this species in some regions indicates a restriction of its historic distribution. In general, the status, movements and habitat requirements are uncertain. They are captured by aviculturists and trapped heavily for food, using decoy individuals, in some areas.

Blue Quail
Caille bleue

Synoicus adansonii

Coturnix adansonii (Verreaux, JP & JBÉ, 1851), Gabon

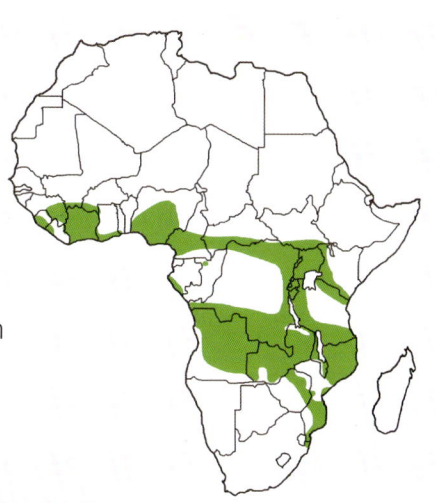

Male Blue Quail from Zimbabwe (photo by Peter Ginn)

Pair of Blue Quails, female on the right, from Uningi Pans, northern Zambia (photo by Derrick Wilby)

Classification

Although closely related to Harlequin Quail *Coturnix delegorguei* and Common Quail *C. coturnix*, their closest relative is the Chinese Painted, Blue-breasted or Asian Blue Quail *Synoicus chinensis* of south-eastern Asia. The African Blue Quail, formerly placed in the genus *Excalfactoria*, is monotypic with no subspecies recognised.

Description

- This is the smallest quail in Africa.
- They are sexually dimorphic.
- Males have a black-and-white throat, a brown back and rufous upper wings, which are distinctive in flight, and a slate-blue breast and belly. The central throat is black, with bold white moustache patches and an extended white bib. They do not have the white stripe above the eye of the male Harlequin Quail and, although less conspicuous, of the Common Quail.
- Females are pale brown, with a streaked back and barred breast and flanks. They are difficult to distinguish from female Common Quail and Harlequin Quail, except for the barred, rather than streaked, breast and flanks.
- Both sexes have bright yellow legs and feet.
- Males have a black bill while the female's bill is pale brown.
- Juveniles resemble adult females, but have pale shaft streaks on the rump, upper tail and wings.
- Downy chicks have brown upperparts with a buff crown streak and three strong, darker longitudinal streaks on the back that converge with one another on the rump.

Distribution and habitat

- They occur widely across Africa, from Sierra Leone and southern Mali, eastwards to western Kenya, and southwards to south-central Angola, Zimbabwe and Mozambique.
- They are nomadic residents in West Africa.
- Outside of forests and dry areas, they are widespread across southern Gabon, southern DRC, north-western Angola and Zambia.
- They are rare in southern Africa, confined to north-eastern Zimbabwe and the coastal plain of Mozambique. Although occasionally recorded during winter in Zimbabwe, usually following above-average summer rains, they are largely a summer nomadic visitor to southern Africa.
- They prefer dense, grassy habitats, such as thick grass along watercourses and around swamps, the edges of cultivated lands and rice fields.

- They usually occur in wetter flooded grasslands and in more wooded areas than sites frequented by Harlequin Quail.
- Prolonged droughts and the degradation of wetlands and grasslands may impact the occurrence of this quail.

Habits

They are seldom as numerous as Common Quail or Harlequin Quail, usually occurring singly, in pairs or small, loose groups. They are reluctant to flush and fly direct and fast for a short distance (20–40 m) when flushed. They are seen most often in the early morning when they emerge into the open to feed. Their call is a piping, three-note whistle, '*ki-kew-yu*', with the first note being the loudest. They are less vocal than the other two African quails. The flush call is also high pitched and three-noted.

Food and feeding

Their diet consists of the seeds of grasses and weeds, and some green plant parts, as well as invertebrates, including termites. The availability of preferred food types may be determined by good rainfall, since local movement often follows rainfall patterns.

Breeding

Breeding in southern Africa is mostly during January to April and in East and West Africa during April to July. The nest is a grass-lined scrape within dense sedges or grass. The clutch of 6–9 thick-shelled eggs are pale yellowish-brown to olive and measure 24.9 x 19.5 mm. Incubation is probably by the female and lasts for 16 days.

Status and conservation

Concern about their stability is associated with increasing transformation and degradation of African wetlands. The only modification of wetlands which favours this species is the establishment of rice paddies. The Asian Blue Quail is a popular aviary bird, but the conservation implications of that species interbreeding with local wild African Blue Quail are probably minimal because of the small chance of any escapees of domesticated foreign birds surviving in the wild.

PART 5

Sandgrouse

SANDGROUSE

Classification

Sandgrouse form the family Pteroclidae, consisting of 16 species distributed across the arid areas of Africa, south-western Europe, and central and southern Asia. They have been associated variously with terrestrial gamebirds (Galliformes), pigeons (Columbiformes) and waders (Charadriiformes). There are two sandgrouse genera:

- *Pterocles* Temminck, 1815, with 14 species, all occurring in warm to hot deserts.
- *Syrrhaptes* Illiger, 1811, which contains two species, both confined largely to cold deserts. The fused three forward toes that characterise the two *Syrrhaptes* species form a 'sand-shoe' that may help the birds walk more easily through sand. *Syrrhaptes* species also lack a hallux (fourth, hind toe), which is typically present in other birds, including *Pterocles*.

Description

Their short bill and legs, seed-eating habits and overall external appearance are all reminiscent of gamebirds. Linnaeus initially assigned sandgrouse to the genus *Tetrao*, which they shared with a range of grouse and spurfowls. However, their long, pointed wings are like those of some pigeons, which also have a short bill and legs, and feed predominantly on seeds.

Natural history

Of the 14 *Pterocles* species, six are endemic to Africa, six are shared between northern Africa and Asia, while India and Madagascar each have one endemic species. All *Pterocles* species have a thick skin, specialised foot sole scales and densely packed feathers with thick underdown, which help them to cope with the high temperatures of deserts. A lower-than-usual (based on body size) metabolic rate further helps to reduce the production of body heat, which is another way they have adapted to cope with a hot environment. The most charismatic desert-related adaptation of sandgrouse is the capacity of their belly feathers to absorb water. During the breeding season, male sandgrouse soak up water in this way to transport the water to flightless chicks in the desert. 'Belly-wetting' behaviour can be an indicator of whether the chicks in a population have begun hatching and of what proportion of the males at a watering site are tending chicks. Chicks are precocial, leaving the nest and feeding themselves on seeds soon after hatching. However, they rely on the female for shelter from the sun,

PREVIOUS PAGE A leopard ambushes Burchell's and Namaqua Sandgrouse at a drinking site in the Kalahari, South Africa (photo by Matt Prophet)

and for the supply of drinking water by the male, for up to 6 weeks before they can fly and are independent of their parents.

Sandgrouse can be choosy about where they drink, preferring sites which are less suitable for ambush by raptors, and generally avoiding relatively saline water. They are dependent on water, often flying long distances daily to water sources. They drink by dipping the bill into the water, taking a couple of sips and tilting the head back to swallow. Sandgrouse are also adapted to exploit the often ephemeral availability of huge quantities of protein-rich seeds produced by annual plants within semi-arid ecosystems, and from hatching they feed exclusively on hard seeds, especially those of legumes (Fabaceae). Their oesophagus is enlarged to form a crop, within which thousands of tiny seeds may be stored. Grit is also eaten to assist with grinding up the seeds in the gizzard. The only other group of birds that feed exclusively on seeds from the time of hatching are the seed-snipes of South America, members of the wader family Thinocoridae.

Sandgrouse are nomadic or migratory and often gregarious. They can occur at high densities, particularly when they gather daily to visit their traditional drinking sites. However, once they take to the ground, sandgrouse can be difficult to find. They appear to actively seek soil types that allow them to blend into the background, and tend to skulk within patchy vegetation rather than seek open ground. Adults may perform injury-feigning distraction displays near eggs or young, giving a distraction call.

Evolutionary placement

- The appropriate evolutionary placement for the sandgrouse was settled in the mid-1960s by South African ornithologist Gordon Maclean, who noted that sandgrouse calling behaviour, nest architecture, clutch size, egg pigmentation, as well as chick plumage and development suggest a closer, albeit distant, relationship to the waders (Charadriiformes).
- Also, despite their short legs, sandgrouse can move around quite well on foot, another feature that links them with waders and not pigeons.
- This was subsequently supported by studies of their DNA, which suggested that they diverged evolutionarily from a wader-like ancestor about 80 million years ago.
- However, more recently comprehensive research on their DNA places them back among the pigeons.
- Despite this, sandgrouse are so distinct in terms of their anatomy, behaviour and genetics, that they are best placed in their own order, Pterocliformes.

Conservation

Sandgrouse have been exploited for food by humans for centuries. In southern Africa, Kalahari tribesmen prevented sandgrouse access to isolated watering sites until they dropped exhausted to the ground. African sandgrouse have also provided wing-shooters with sporting quarry for more than a century. In 1902, in his book *Bird Hunting on the White Nile*, wing-shooter Harry Witherby writes: 'Sandgrouse [in this case Chestnut-bellied Sandgrouse *P. exustus* and Spotted Sandgrouse *P. senegallus*] shooting along the banks of the White Nile afford such sport that millionaires would give untold gold for similar shooting were it to be had in England or Scotland.' More recently, commercial wing-shooting of Namaqua Sandgrouse *P. namaqua* and Burchell's Sandgrouse *P. burchelli* has been offered in the Northern Cape province of South Africa. This wing-shooting 'industry' significantly supplements the incomes of farmers in that semi-arid part of Africa.

Their occurrence in generally remote and relatively inhospitable environments render them largely free of any major threats which therefore allows the conservation status of all of the African sandgrouse to be listed as Least Concern (The IUCN Red List of Threatened Species v 2015-4).

Pin-tailed Sandgrouse
Ganga cata

Pterocles alchata

Tetrao alchata (Linnaeus, 1766), Spain

LEFT Male Pin-tailed Sandgrouse showing the pure white belly and underwing in flight (photo by Francois Mougeot)

BELOW LEFT Incubating male Pin-tailed Sandgrouse with a research telemetry transmitter (photo by Francois Mougeot)

BELOW Pair of Pin-tailed Sandgrouse, female on the left (photo by Francois Mougeot)

> ## Classification
>
> There are two subspecies recognised:
>
> *P. a. alchata* (Linnaeus, 1766) in the Iberian Peninsula and south-eastern France
>
> *P. a. caudacutus* (Gmelin, SG, 1774) in northern Africa through the Middle East to southern Kazakhstan and Pakistan

Description

- These are the only sandgrouse with a pure white belly and underwing.
- They have long needle-like tail feathers, similar to Spotted Sandgrouse *P. senegallus* and Chestnut-bellied Sandgrouse *P. exustus*.
- In flight, both sexes show a distinctive white wing-bar and white underwing with a black tip.
- Males have greenish-yellow upperparts mottled with yellow, orange-buff sides of the head, a black line through the eye and a maroon breast band and black throat.
- Females have a pale buff breast with three black bands and a distinctive white throat.
- Juveniles have black and buff concentrically marked upperparts and breast.
- Downy chicks have a rust-coloured crown with black edges and a whitish line down the mid-crown.
- The back has four whitish and rust-brown patches that form a double figure-eight pattern.

Distribution and habitat

- They are resident and nomadic north of the high Atlas Mountains in Morocco, Algeria, Tunisia and Libya, except the band of north-eastern plains.
- They inhabit arid and semi-arid steppes, including sparsely vegetated stony ground, dry plains and plateaus, mudflats and dry coastal river beds and depressions (*wadis*), but avoid open deserts.

Habits

They are gregarious even roosting as a flock. They drink in the early morning, 1–2 hours after sunrise. They rest in the heat of the day and gather to dust-bathe in the evening. Their flight call is a loud far-reaching ringing '*kata, kata*' uttered frequently.

Food and feeding

Their diet consists almost exclusively of small seeds, mostly of legumes and *Polygonum*, *Fagopyrum*, *Salicornia*, *Artemisia*, *Alhagi*, *Helianthemum* and *Asphodelus* species. They move onto farmlands in the early boreal summer to feed on wheat, oat grains, lentils and other cereals.

Breeding

They are monogamous and mostly solitary nesters, but sometimes nest in loose colonies. The breeding season is mostly from April to June, after the rains. The nest is a scrape sometimes close to a large stone or bush. The clutch of 2–4 long elliptical smooth eggs are slightly glossy buff, variably spotted, blotched or speckled with brown and grey. They measure 47.0 x 30.0 mm. Incubation by the male during the day and female during the night lasts for about 21 days.

Status and conservation

Their populations are suspected to be stable in the absence of evidence for any declines or substantial threats. They are common in Algeria and Morocco, except for western Morocco, where numbers have declined in recent decades.

Namaqua Sandgrouse
Ganga namaqua

Pterocles namaqua

Tetrao namaqua (Gmelin, JF, 1789) Namaqualand, Northern Cape Province, South Africa

Male Namaqua Sandgrouse in flight
(photo by Ian White)

Female Namaqua Sandgrouse
(photo by Eugene Vorster)

Incubating male Namaqua Sandgrouse
(photo by Penn Lloyd)

Pair of Namaqua Sandgrouse
with cryptic chicks
(photo by Penn Lloyd)

Classification

They are most closely related to Black-bellied Sandgrouse *P. orientalis* and Chestnut-bellied Sandgrouse *P. exustus*. Namaqua Sandgrouse are monotypic with no subspecies recognised.

Description

- They have long, pointed central tail feathers.
- Males have a plain olive-yellow head and neck, a white-and-chestnut breast band and a dark brown belly.
- Females have a yellow face and throat, mottled pale orange and grey on brown on the back and wings, barred on the belly and streaked on the breast.
- Juveniles resemble adult females but their back plumage is tinged with rufous, with darker barring.
- Each feather has a whitish border and the tail lacks the long central feathers.
- Downy chicks have a yellowish-brown crown and nape with irregular white bars, edged with black.
- The face is light brown, with a whitish stripe in front of the eye, which trails off down to the throat.
- The upperparts are also yellowish-brown, strongly marked with irregular white stripes and bars.

Distribution and habitat

- They are a southern African near-endemic, with a minimal extension into southern Angola.
- North-western populations are nomadic, while south-eastern populations are migratory, concentrating in the Northern Cape, South Africa, during the breeding season.
- During the austral winter non-breeding season, these populations migrate to the north and east into the southern Kalahari.
- They inhabit stone and gravel deserts and open semi-deserts with a sparse scattering of low shrubs or grass tufts, and occasionally arid sandy savanna with denser vegetation.

Habits

Flocks drink 2–3 hours after sunrise. In the evening, flocks fly to roosting sites in stony areas, where each bird makes a shallow roosting scrape. Favoured drinking sites have clear, fresh water with low salinity, gently sloping, open banks and limited tall vegetation, which might provide cover for predators, particularly raptors. They land several metres from the water, where they socialise and dust-bathe, then run down to the water to drink and belly-wet. They have a nasal, three-note flight call, '*ki-ki-vee*' or '*kel-kie-wyn*'. The take-off call is a rapid '*kip-kip-kip-kip*' and flocks mutter '*kip*' notes continuously while on the ground at water.

Food and feeding

They feed on the seeds of annual plants, including *Indigofera* species in the Nama Karoo, *Lotononis* species in the Succulent Karoo, *Tephrosia* species in Namibia and *Tephrosia* species and *Requernia sphaerosperma* in the southern Kalahari. Feeding birds walk slowly with head bent, pecking rapidly at the ground. They do not scratch with their feet, but flick the sand with their bill.

Breeding

They are monogamous solitary nesters. Namibian populations breed from January to August, peaking in May. South African populations breed from June to November in the Kalahari, August to January in the Nama Karoo, and September to February in Namaqualand and the south-western Western Cape. They nest on sparsely vegetated calcrete and sandy flats. The nest is a scrape in the sand, often placed next to stones or a tuft of grass, which act as disruptive camouflage. They have a 3-egg clutch, with each egg measuring 36.1 x 25.2 mm. The eggs are variable in appearance between clutches and in some cases within clutches, some being sparsely speckled with a light pigment over a cream background, while others are covered with heavy, dark speckles on a dark beige background, and still others have the speckling concentrated in a band around the thickest part of the egg. Females incubate during the day, while males incubate at night. Incubation lasts 21 days.

Status and conservation

Although there is no evidence of extensive changes in their distribution or numbers, large flocks at traditional watering sites in the Northern Cape have declined since the late 1940s. This may be due to increased dispersion in response to an increased number of artificial watering points. Nest predation is primarily by Yellow Mongoose *Cynictis penicillata* and Common Egg-eater *Dasypeltis scabra* snakes. Telephone lines and fences located in the flight path to and from favoured watering sites cause fatal collisions. Wing-shooting should only be during the non-breeding months.

Chestnut-bellied Sandgrouse
Ganga à ventre brun

Pterocles exustus

Pterocles exustus Temminck, 1825, Senegal

LEFT Male Chestnut-bellied Sandgrouse from Nigeria (photo by Tasso Leventis)

BELOW Two female and a male Chestnut-bellied Sandgrouse (photo by Tasso Leventis)

GAMEBIRDS

ABOVE A male and two female Chestnut-bellied Sandgrouse flush from a drinking site (photo by Peter W Hills)

TOP RIGHT Male Chestnut-bellied Sandgrouse belly-wetting to carry water to his flightless chicks (photo by Tasso Leventis)

BOTTOM RIGHT A male Chestnut-bellied Sandgrouse scans the sky above for potential danger (photo by Tasso Leventis)

Classification

They are most closely related to Black-bellied Sandgrouse *P. orientalis* and Namaqua Sandgrouse *P. namaqua*. There are six subspecies recognised:

P. e. floweri Nicoll, 1821, in Egypt

P. e. exustus Temminck, 1825, in Mauritania, Senegal and Gambia to southern Sudan

P. e. ellioti Bogdanov, 1881, in south-eastern Sudan, Eritrea, northern Ethiopia and Somalia

P. e. olivascens (Hartert, EJO, 1909) in south-western Ethiopia, Kenya and northern Tanzania

P. e. erlangeri (Neumann, 1909) in the western and southern Arabian Peninsula

P. e. hindustan Meinertzhagen, R, 1923, in south-eastern Iran, Pakistan and India.

Description

- They have long, needle-like tail feathers, similar to Namaqua Sandgrouse, Spotted Sandgrouse *P. senegallus* and Pin-tailed Sandgrouse *P. alchata*.
- Males have a distinctive yellow head and wings, and although the belly and underwing appear black it is actually dark chestnut.
- They have a narrow black breast-band, without black on the throat, which differentiates from Black-bellied Sandgrouse.
- Females have dark brown barred upperparts and belly, and the breast is flecked, not plain.
- Juveniles have yellowish upperparts, throat and breast with fine dark vermiculations, and no breast band.
- Downy chicks have a dark streak above the bill separated from the lores by a creamy white line, a golden buff patch around the eye which is edged black.
- They have a white line down the back, flanked by golden buff patch lines edged with black.

Distribution and habitat

- Their distribution stretches in a band from the Mauritanian coast across central West Africa to the Red Sea and Gulf of Aden in East Africa, along the lower Nile River basin in Egypt, and through Arabia to India.
- They prefer open semi-desert and sandy arid scrub, as well as cultivated or fallow fields and grasslands.

Habits

They are gregarious gathering in large flocks to drink 2-3 hours after sunrise, and in the evening during hot weather. They approach the waterhole from height, circling and landing 20-30 m from the water. When disturbed, they crouch or creep silently away until hidden. They have a harmonious deep-toned '*kwit-gurut, kwit kwit-gurut*' flight-call and mutter a musical '*creen*' murmur at the waterhole.

Food and feeding

They forage mostly during the cooler morning and late evening. Their diet consists mostly of hard legume and grain seeds, *Trianthema*, *Indigofera* and *Heliotropium* species, as well as some plant shoots and insects, including ants.

Breeding

They are monogamous, solitary nesters. They breed during March to July in the west and during April to June in the east. The nest is a shallow unlined scrape in open desert. The clutch of three elongate, equally rounded eggs are creamy buff, greyish to light olive, heavily spotted and blotched with brown and pale purplish grey, and measure 36.1 x 25.2 mm. Incubation by the female during the day and by the male at night lasts 22–23 days.

Status and conservation

They are locally abundant, with large flocks in north-central Mali, Sudan, Somalia and the lowlands of Ethiopia, and are the commonest sandgrouse in central Chad.

Spotted Sandgrouse
Ganga tacheté

Pterocles senegallus

Tetrao senegallus (Linnaeus, 1771), Algeria

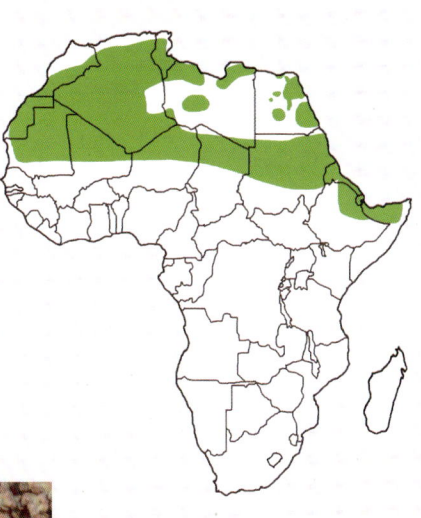

LEFT A flock of Spotted Sandgrouse at a drinking site (photo by Nik Borrow)

BELOW LEFT Female Spotted Sandgrouse (photo by Nik Borrow)

BELOW Male Spotted Sandgrouse (photo by Nik Borrow)

Classification

They are monotypic with no subspecies recognised.

Description

- They are a rather pale yellowish-buff sandgrouse, with elongated needle-like tail feathers and less black on the belly than the short-tailed Black-bellied Sandgrouse *P. orientalis*.
- Males have mostly uniform upperparts, no black on the face or throat, an orange-yellow throat and a blue-grey crown and hind neck which meets the blue-grey breast band.
- In flight, they show black secondaries and a distinctive black line down the mid-belly.
- Females have spotted upperparts and a breast band.
- Juveniles have sandy upperparts with darker crescentic bars and somewhat vermiculated streaks.
- Downy chicks are light greyish brown with obscure mottling.

Distribution and habitat

- They are distributed in North Africa, Israel and Arabia to Afghanistan and Pakistan.
- In Africa, they are spread from southern Morocco and Mauritania across to Egypt, Eritrea and northern Somalia.
- They inhabit a wide range of dry open country from flat stony desert plains (*hammadas*) to sandy desert with isolated patches of vegetation.

Habits

They are resident, nomadic or partially migratory. Flocks of up to 60 individuals circle and land close to a waterhole each morning, and less so again in the evening during hot weather. They are very vocal and vigilant, approaching the water to drink and flushing at any disturbance. They roost in loose flocks, with each bird making a shallow roost scrape. Their call is a distinctive, liquid and musical '*cuito, cuito*' or '*wittoo, wittoo*'.

Food and feeding

They feed largely on small hard seeds, including those of *Euphorbia guyoniana* and *Asphodela tenuifolia*. In Egypt, they gather with Crowned Sandgrouse *P. coronatus* to feed on grain spilled from trucks transporting between Red Sea ports and Nile valley markets.

Breeding

They are monogamous solitary nesters, breeding mostly from March to July. The nest is a small unlined scrape on flat ground, sometimes placed among stones about the same size as the bird. The clutch averages 3 eggs, which are elliptical and equally rounded at each end. They are buff, sparingly spotted and scrawled with light brown, reddish brown and purplish grey, and measure 41.7 x 28.0 mm. Incubation by the female during the day and by the male during the night lasts for 29–31 days.

Status and conservation

They are common to abundant in most parts of their range, except Morocco where their distribution is patchy and they are less common.

Black-bellied Sandgrouse
Ganga unibande
Pterocles orientalis

Tetrao orientalis (Linnaeus, 1758), 'In Orient' = Anatolia, ex Hasselquist

Female Black-bellied Sandgrouse (photo by Tasso Leventis)

Male Black-bellied Sandgrouse (photo by Tasso Leventis)

Black-bellied Sandgrouse at a drinking site (photo by Sergey Dereliev)

Classification

They are most closely related to Chestnut-bellied Sandgrouse *P. exustus* and Namaqua Sandgrouse *P. namaqua*. There are two subspecies recognised:

P. o. orientalis (Linnaeus, 1758) in the Canary Islands, Morocco to Libya, and the Iberian Peninsula to western Iran

P. o. arenaruis (Pallas, 1775) in Kazakhstan to southern Iran and Afghanistan east to north-western China

Description

- They are a large sandgrouse with the average male weighing 514 g while the female weighs about 434 g.
- Their short tail, completely black belly and black-and-white underwing coverts are distinctive.
- Males are distinguished from Spotted Sandgrouse *P. senegallus* by a black throat, black band across the breast and contrasting yellow wings.
- Females have a dark band below their breast spots and an extensive black belly.
- They also have a pale throat, buff in the male and white in the female.
- Juveniles are similar to adults, but greyer.
- Downy chicks are ochre-tawny above with black tips to the down.
- They have a bold cross-shaped pattern of buffy white lines dividing the tawny back into a bold double figure-eight pattern.

Distribution and habitat

- They occur in north-western Africa from Morocco to Libya, the Canary Islands, southern Iberia, Cyprus, and the Middle East to south-central Asia.
- They are also a regular non-breeding migrant to Egypt.
- In north-western Africa, they prefer the semi-arid steppe north of the Atlas Mountains.
- They avoid open desert.

Habits

They are sedentary or nomadic in north-western Africa. Flocks drink at watering holes with no vegetation on the banks between 1–2 hours after sunrise. They are reluctant to fly during the heat of the day, often seeking shade. Their call is a gruff-mellow '*churr-rur-rur*' or '*tchourou*'.

Food and feeding

Their diet consists mostly of small seeds with a preference for legume species. They will also eat some agricultural grains and insects including termites, beetles and their larvae.

Breeding

They are monogamous solitary nesters. Breeding occurs from April to July in northwestern Africa. The nest is a shallow unlined scrape. The clutch of 3 elliptical eggs are pale to deep buff or greyish, smudged, blotched and spotted with brown and grey, which measure 47.0 x 32.0 mm. Incubation by the female during the day and by the male at night lasts 21–22 days.

Status and conservation

They are common in Algeria and Libya, scarce in Tunisia, and absent from the northeast.

Yellow-throated Sandgrouse
Ganga à gorge jaune

Pterocles gutturalis

Pterocles gutturalis Smith, A, 1836, Kurrichane (Zeerust), North West Province, South Africa

Male Yellow-throated Sandgrouse subspecies *P. g. saturatior* from Kenya (photo by Jacques Pitteloud)

Female Yellow-throated Sandgrouse subspecies *P. g. saturatior* from Kenya (photo by Jacques Pitteloud)

Pair of Yellow-throated Sandgrouse subspecies *P. g. gutturalis* from southern Africa (photo by Maans Booysen)

Classification

Two subspecies are recognised:

P. g. gutturalis Smith, A, 1836 in South Africa, north-eastern Namibia, eastern and northern Botswana, western Zimbabwe and Zambia

P. g. saturatior Hartert, EJO, 1900 in Ethiopia, Kenya, Tanzania and north-eastern Zambia

The subspecies *P. g. tanganjicae* described for the population in Tanzania is not widely accepted.

Description

- They are a large sandgrouse with distinctive dark chestnut belly feathers that are conspicuous in flight, dark brown underwing plumage and a wedge-shaped tail.
- Both sexes have blue-grey, bare-skinned eye rings.
- Males have a pale creamy-yellow face and throat encircled by a black band, and a dull grey breast.
- Females also have a yellowish throat, but lack the black collar, and the back and breast are mottled with black and brown.
- Juveniles are similar to adult females, but the back is barred olive-buff and the mottling is finer.
- Downy chicks have a strongly marked crown and back with a network of rufous-brown patches, edged with black and separated by a matrix of white lines in a double figure-eight pattern.
- The sides of the head are dark buff, with a buffy-white stripe above the eye.
- The underparts are greyish-white, washed with brown on the breast.

Distribution and habitat

- They have a widespread but fragmented distribution throughout southern and eastern Africa, ranging from South Africa up through Zambia, Tanzania, Kenya and Ethiopia.
- They are variably resident or nomadic with some sub-regional migration.
- They prefer more mesic habitats than most sandgrouse, commonly frequenting short, open grasslands on deep, alluvial soils, often near swamps and rivers.
- In agricultural areas they will also forage in recently burnt grasslands and cultivated fields.

Habits

They drink during the mid-morning, and tolerate more shrubs and trees in the near vicinity of a watering site than open-country sandgrouse. They are found in flocks of 2 to 20 birds, but migratory flocks of up to a couple of hundred are sometimes encountered. When disturbed, they freeze before flushing with heavy, audible wing beats. The flight call is a loud, guttural, two-syllable '*aw aw*'. The take-off call is a harsh '*glock-glock-glock*'.

Food and feeding

Because they inhabit relatively moist habitats, they have a diet of more varied seed types than dry-country sandgrouse. They also use manipulated habitats, often feeding on the seeds of pioneer plants including *Crotalaria*, *Cassia*, *Sesbania* and *Indigofera* species, and weed plants such as *Hibiscus*, *Helianthus*, *Amaranthus*, *Bidens* and *Crotalaria* species. They will even take advantage of waste cereal grains such as oats, wheat, barley and sorghum.

Breeding

They are monogamous solitary nesters, breeding during the drier months from May to September, with a peak in June. The nest is a simple scrape, usually lightly lined with twigs and dry grass placed in a fairly well vegetated site, particularly where weedy fields are used. The 3 eggs are dusky-brown or tawny, with lines and blotches of umber occasionally forming a broad ring around the thickest part, and measure 45.8 x 33.9 mm. The female incubates by day and the male at night. The incubation period of 26–27 days is relatively long compared to that of smaller sandgrouse. When an incubating bird is approached, it first crouches on the nest and then walks off some distance before flushing.

Status and conservation

Their use of agricultural lands may expose them to the misuse of herbicides and to nest losses when these lands are ploughed during the late austral winter in preparation for planting. Nest predators include Banded Mongoose *Mungus mungo* and Pied Crow *Corvus albus*. Hunting may be detrimental where population levels may not be sufficient to allow for a sustainable off-take.

Crowned Sandgrouse
Ganga couronné

Pterocles coronatus

Pterocles coronatus Lichtenstein, MHC, 1823, Nubia

Crowned Sandgrouse
(photos by Tasso Leventis)

Classification

There are five subspecies recognised:

P. c. coronatus Lichtenstein, MHC, 1823 in the western and central Sahara to eastern Egypt and northern Sudan

P. c. vastitas Meinertzhagen, R, 1928 in the Sinai Peninsula and north-eastern Egypt to Israel and Jordan

P. c. saturates Kinnear, 1927 in northern Oman

P. c. atratus Hartert, 1902 in the Arabian Peninsula to southern Afghanistan

P. c. ladas Koelz, 1954 in Pakistan

Description

- They are a relatively small (average mass 300 g) short-tailed sandgrouse.
- Males have a white forehead bordered by a black stripe from the base of the bill to the front of the crown.
- The sides of the face are white, the crown is pale rufous with a pale blue-grey stripe from above and behind the eye to the nape.
- The chin and upper throat are black with ochre-gold on the lower throat, cheeks and hindneck.
- The back is rufous sandy.
- Females are similar to the female Lichtenstein's Sandgrouse *P. lichtensteinii*, but have an unspotted yellow throat.
- Juveniles are similar to adult females but are more rufous-buff with coarser barring and a whitish throat.
- Downy chicks are rufous-buff above with a similar pattern to that of the downy chick of Namaqua Sandgrouse *P. namaqua*.

Distribution and habitat

- They are widespread across the Sahara from Morocco to the Nile River and on the Red Sea coast in Egypt and Sudan.
- They inhabit widespread hot, arid habitats from stony deserts to mountainsides.

Habits

They are not as gregarious as other sandgrouse. Flocks fly fast to water in the early morning, and not like many other sandgrouse will drink quite brackish water. The '*cla-cla-cla*' or '*cha-chagarra*' chattering call is high pitched and rises in pitch.

Food and feeding

Their diet consists almost exclusively of small hard seeds, particularly those of *Tephrosia apollinea* and *Asphodela tenuifolia*. They will also feed on spilled grains along roads.

Breeding

They are monogamous solitary nesters. Breeding is from May to July in Morocco and Algeria, and during March in Chad. The nest is an unlined scrape or natural hollow, sometimes with a ring of small stones round the rim. The clutch of 3 elliptical smooth eggs are cream to pale yellowish, evenly spotted and blotched with light brown and pale purplish grey. They measure 39.4 x 27.4 mm. Incubation is by the female during the day and by the male during the night. The incubation period is unrecorded.

Status and conservation

They are uncommon in Morocco, otherwise locally common and are the most common sandgrouse in the Ksours Mountains of Algeria.

Black-faced Sandgrouse
Ganga à face noire

Pterocles decoratus

Pterocles decoratus Cabanis, 1868, Lake Jipe, Kenya

Pair of Black-faced Sandgrouse, male on the right (photo by Jacques Pitteloud)

Female Black-faced Sandgrouse (photo by Jacques Pitteloud)

Male Black-faced Sandgrouse (photo by Jacques Pitteloud)

Male Black-faced Sandgrouse calling (photo by Nik Borrow)

Pair of Black-faced Sandgrouse at a watering site in Ethiopia (photo by Peter W. Hills)

Classification

There are three subspecies recognised:

P. d. ellenbecki Erlanger, 1905 in southern Ethiopia and southern Somalia to northern Kenya and north-eastern Uganda

P. d. decoratus Cabanis, 1868 in south-eastern Kenya and eastern Tanzania

P. d. loveridgei (Friedmann, 1928) in south-western Kenya and central Tanzania

Description

- They are a relatively small sandgrouse with a short tail and a black belly.
- Males have a black face patch from the fore crown down onto the throat and narrow black-and-white stripes curving backwards from above and behind the eyes.
- The narrow black breast band and broad white band across the upper belly distinguishes both sexes from all other sympatric sandgrouse.
- Lichtenstein's Sandgrouse *P. lichtensteinii* also have a black-and-white face pattern, but have no black on the throat and no pronounced supercilium.
- Juveniles are similar to adult females but have rusty upperparts with narrower, darker barring.
- Downy chicks are sandy to gold and grey above, with bold black mottling.

Distribution and habitat

- They are distributed across southern Ethiopia, southern Somalia, north-eastern Uganda, Kenya and in eastern and central Tanzania.
- They are found in dry open savanna, thorn bush, semi-desert scrub and coastal dunes, often seen in bare areas, road verges and in small open spaces in thick bush.

Habits

They are thought to be mostly sedentary. Mostly encountered singly or in pairs, they gather in flocks to drink. They usually drink 1–4 hours after sunrise, and sometimes at dusk. They frequently flock together with Chestnut-bellied Sandgrouse *P. exustus* in East Africa. Their flight-call is a repeated three-note '*chuck-chuck-chuk*' or '*wop dela wiiii*' whistle of two long syllables and one short syllable, and a wispy '*tseeoo whit-i-weeer whit-i-weeer*' flush call.

Food and feeding

Their diet consists mostly of seeds, mainly legumes, including *Trianthema salsoloides*, *Indigofera* species and *Heliotropium undulatifolium*.

Breeding

They are monogamous solitary nesters. They breed mostly from June to August. The nest is an unlined scrape on bare sandy or stony ground. The clutch of 3 long, oval, rounded greyish buff eggs, spotted or blotched with reddish brown and mauve-grey, measure 33.5 x 25.5 mm.

Status and conservation

They are not threatened globally and are locally common throughout their distribution.

Lichtenstein's Sandgrouse
Ganga de Lichtenstein
Pterocles lichtensteinii

Pterocles lichtensteinii Temminck, 1825, Nubia

Male Lichtenstein's Sandgrouse (photo by Nik Borrow)

Female Lichtenstein's Sandgrouse (photo by Nik Borrow)

Classification

There are five subspecies recognised:

P. l. targius Geyr von Schweppenburg, 1916 in Morocco and Mauritania east to Chad

P. l. lichtensteinii Temminck, 1825 in north-eastern Chad, northern Sudan to Eritrea and northern Somalia, eastern and north-eastern Egypt to Jordan

P. l. sukensis Neumann, 1909 in south-eastern Sudan, southern Ethiopia to central Kenya

P. l. arabicus Neumann, 1909 in the southern Arabian Peninsula, through southern Iran and Afghanistan to Pakistan

P. l. ingramsi Bates & Kinnear, 1937 in Hadhramaut, southern Yemen

Description

- These small short-tailed sandgrouse appear dark because of their close barring.
- Males have a distinctive black-and-white marked crown and forehead, a white supercilium extending backwards, yellow breast with two black breast bands, an orange and a yellow breast band, and heavily barred neck and wing coverts.
- Females are uniformly barred, lacking any distinctive markings, similar to female Crowned Sandgrouse *P. coronatus*, but their face and throat are spotted, not plain yellow.
- Juveniles are similar to adult females, but are more closely barred above and below.
- Downy chicks are brown with light chocolate brown lores, supercilium, ear coverts and below the eyes, bordered below by a pale malar stripe and above by a pale lateral crown-stripe running from the bill to the nape.

Distribution and habitat

- They are distributed across northern and eastern Africa, southern Jordan, Arabia, Socotra, Iran and Pakistan.
- In Africa, they occur in southern Morocco, Mauritania, Senegambia, eastern Mali, adjacent Niger, southern Algeria, northern and eastern Chad, Sudan, south-eastern Egypt, Ethiopia, northern Somalia and Kenya.
- They inhabit rocky and bushy desert areas and arid bushveld with scattered shrubs, avoiding open desert.

Habits

They are less gregarious than other sandgrouse, and fly singly just above vegetation to drink after dark and before dawn. They are selective of drinking sites with low salinity water. They spend most of the day inactive in the shade of a rock or bush. Mostly silent in flight, but will utter a double liquid-whistle '*whittou-whittou*' call occasionally.

Food and feeding

Their diet consists largely of the seeds of *Vachellia* (*Acacia*) *seyal*, '*A*'. *ehrenbergiana* as well as *Asphodelus*, *Cassia* and *Prospia* species. They will also eat the flowers and leaves of *Mesembryanthemum* species, and some insects (beetles, ants, ant-lions) and their larvae.

Breeding

They are monogamous solitary nesters, breeding mostly from May to July. The nest is an unlined scrape usually among scattered trees or rocks. The clutch of 2–3 elliptical olive, brownish-olive, pinkish or very pale buff eggs are blotched, spotted or sparsely marked with reddish brown and underlying purplish grey. They measure 42.0 x 26.3 mm. Their incubation biology is unknown.

Status and conservation

Their global population size is alleged to be stable in the absence of evidence for any declines or substantial threats, and they are generally locally common, although uncommon in West Africa, Sudan and Kenya.

Four-banded Sandgrouse
Ganga quadribande
Pterocles quadricinctus

Pterocles quadricinctus Temminck, 1815, Senegal

ABOVE Female Four-banded Sandgrouse in Nigeria (photo by Tasso Leventis)

LEFT Male Four-banded Sandgrouse in Cameroon (photo by Tasso Leventis)

> ## Classification
> This is a monotypic species with no subspecies recognised.

Description

- This is a smallish short-tailed sandgrouse.
- Males have a chestnut, white, and black chest band, and a black-and-white forehead pattern similar to that of Double-banded Sandgrouse *P. bicinctus*, Black-faced Sandgrouse *P. decoratus* and Lichtenstein's Sandgrouse *P. lichtensteinii*, but are not sympatric with the former and lack the black throat of *P. decoratus* and barred throat and breast of *P. lichtensteinii*.
- They also have a plain buff-brown neck and upper breast.
- In flight, blackish flight feathers contrast with grey underwing coverts.
- Females have a plain buff breast and barred belly.
- Juveniles are more rufous than adults with finer black barring.
- Downy chicks have a dorsal pattern similar to most sandgrouse chicks, but with fewer white lines creating a single not double figure-eight pattern on the back.

Distribution and habitat

- They are distributed from Senegal and The Gambia across the Afrotropics in a narrow band to Eritrea, Ethiopia and western Kenya.
- They inhabit dry, open or partly wooded savannas, open and bushy grasslands, scrubby coastal dunes and cultivated pastures often on clay or stony soils.

Habits

They are usually seen in pairs or small flocks which are sedentary to partial migrant and mostly nocturnal, resting inactive during the day. When flushed, they have a zig-zag flight for a short distance before landing. They fly low to watering sites at sunset, thereafter scattering to feed until late at night. Their flight call is a piercing rhythmic whistled '*wur-wulli*' or '*pirrou-ee*', the last note tremulous.

Food and feeding

Like most sandgrouse, their diet consists mostly of small seeds.

Breeding

They are monogamous solitary nesters, breeding from November to June, but mostly from February to March. The nest is a bare scrape, or lightly lined with grass, on bare stony soil or shingle, or among scatter scrub such as *Bauhinia* trees. The clutch of 2–3 eggs equally rounded at each end are clay-pink or salmon-buff, spotted and flecked with orange-brown or pale brown and underlying mauve. They measure 39.6 x 27.6 mm. Their incubation biology is unknown.

Status and conservation

They are fairly common throughout most of their range, locally common in Senegambia, common in Mali, abundant in northern Ivory Coast during dry season, common in Nigeria and Chad south of 16°N, fairly common in Sudan and apparently uncommon in Kenya.

Double-banded Sandgrouse
Ganga bibande
Pterocles bicinctus

Pterocles bicinctus Temminck, 1815, Great Fish River, south-eastern Namibia

Female Double-banded Sandgrouse (photo by Ian White)

Male Double-banded Sandgrouse (photo by Maans Booysen)

Male Double-banded Sandgrouse (photo by Ian White)

Female Double-banded Sandgrouse (photo by Tasso Leventis)

Male Double-banded Sandgrouse (photo by Eric Stockenstroom)

Female Double-banded Sandgrouse (photo by Maans Booysen)

Male Double-banded Sandgrouse (both photos by Ian White)

Classification

Four subspecies are recognised:

P. b. bicinctus Temminck, 1815 in north-western South Africa, Namibia, and Botswana

P. b. multicolor Hartert, 1908 in north-eastern South Africa and eastern Botswana

P. b. usheri (Benson, 1947) in western Mozambique, Zambia and southern Malawi

P. b. ansorgei (Benson, 1947) in south-western Angola

Description

- They are a medium-sized short-tailed sandgrouse.
- Males are the only sandgrouse in their range with a bold black-and-white band across the forehead, between the eyes and the beak, a lightly speckled crown and a yellow bill.
- Females are similar to female Namaqua Sandgrouse *P. namaqua*, but have a barred, not streaked upper breast and, as in the male, rounded tail feathers.
- Juveniles resemble adult females, but their back plumage is more pinkish-brown and has less barring.
- Their upperparts are speckled and finely mottled with black.
- The throat and upper breast are pinkish-buff, highlighted with light brown, with each feather having a brown bar.
- The crown of downy chicks has two dark brown patches, speckled and bordered with black and bisected and surrounded by white.
- The sides of the head are brown, and there is a white stripe above and two white stripes behind each eye.
- Their back has patches of rufous-brown edged with black and separated by white lines that form a double figure-eight pattern.

Distribution and habitat

- Widespread in southern African, they extend marginally into Angola, Zambia and Malawi.
- They are abundant in northern Botswana, northern Zimbabwe and along the Limpopo River system and are the only sandgrouse species in Mozambique.
- They are largely sedentary, although may move in search of water, especially in the dry season.
- They prefer wooded habitats, favouring *Brachystegia* and mopane woodlands and the savannas of the Okavango Delta region in Botswana, the Etosha Pan in Namibia, and the Limpopo River valley and Limpopo Province lowveld in South Africa.

- In the drier, western areas they prefer rocky hill slopes or gravel plains, often far less vegetated than in the eastern parts of their range.

Habits

They drink at last light, often after sunset. Birds weave through adjacent trees then flutter before dropping to the water's edge. Distances covered between feeding, roosting and watering sites are shorter than those travelled by open country sandgrouse species. They are largely inactive during the day, resting beneath scrubby vegetation. They are also less vocal in flight than other sandgrouse. Their flush call is a harsh *'chuck chuck'*. At the drinking site they give a musical rapid, 7–10 syllable *'pitee YOU'RE iti pur'*. When disturbed, they may give a growling *'churr'*.

Food and feeding

They forage during the cooler hours of the morning and evening, and late on moonlit nights. Their diet consists of the seeds of herbs and annual weeds, including *Requernia*, *Tephrosia*, *Cyperus*, *Bidens* and *Datura* species. When foraging, they search a restricted area particularly thoroughly for scattered seeds.

Breeding

They are monogamous solitary nesters breeding from May to October. The nest is a scrape, sparsely lined with dry vegetation, either well concealed in vegetation or more open. The clutch of 3 pinkish-buff eggs, spotted and streaked with reddish-brown and slate-grey, measure 37.1 x 26.6 mm. Unlike most other sandgrouse, the male incubates during various parts of the day and the female incubates at night. The incubation period is 23–24 days.

Status and conservation

An aspect of concern is where they are hunted as an added quarry during game hunts, since the peak game-hunting season during the austral winter straddles their breeding season. Also of concern is that they are shot when approaching their drinking site at dusk, resulting in birds being prevented from drinking at that site and them going through the night without drinking. This can be resolved by leaving nearby alternative watering sites undisturbed.

Burchell's Sandgrouse
Ganga de Burchell
Pterocles burchelli

Pterocles burchelli Sclater, WL, 1922, near Griquatown, Northern Cape Province, South Africa

Male Burchell's Sandgrouse flushes from a drinking site (photo by Ian White)

Female Burchell's Sandgrouse (photo by Dayne Braine)

GAMEBIRDS

Male Burchell's Sandgrouse (photo by Ian White)

Male Burchell's Sandgrouse approaching a watering site (photo Ian White)

Two pairs of Burchell's Sandgrouse in their arid environment (photo by Andre Botha)

Classification

They are monotypic with no subspecies recognised. In southern Africa, they are sometimes called Spotted Sandgrouse, but this creates confusion with the Spotted Sandgrouse *P. senegallus* of northern Africa.

Description

- They have a salmon-pink breast and belly feathers and brownish-yellow upperparts, with overall bold, white spotting.
- Males have a grey face and throat, and a prominent yellow eye patch, while females have a yellowish face with less prominent yellow skin around the eye.
- They have a stocky build and relatively long legs.
- Juveniles resemble adult females, but the breast is buffy, with each feather having a pale tip above a light brown bar.
- The white spots on the wings are duller, and each feather is barred with buff.
- Downy chicks have a golden-buff forehead grading posteriorly to white and bordered with black.
- The crown is white in the centre, bordered with black and surrounded by light brown speckled with black.
- The face and sides of the head are buffy, with brown-and-white stripes, and there is a light brown ring around the eye.
- The back has a striking white double figure-eight marking, bordered with black.

Distribution and habitat

- They are a near-endemic in southern African, extending marginally into south-eastern Angola and south-western Zambia.
- Their distribution is centred on the sandy soils of the Kalahari basin, from western South Africa and extreme north-western Zimbabwe, across Botswana and the north-eastern half of Namibia, avoiding the driest southern and western regions of Namibia.
- They are mostly found on red Kalahari sands with dry savanna or scattered scrub, or shrub and grass tufts up to 50 cm tall, mostly in arid sweet bushveld and Kalahari thornveld.

Habits

Although largely sedentary, they may be nomadic in response to the availability of seeds. They arrive to drink 2–3 hours after sunrise, landing directly at the water's edge or even in the shallow water, drink rapidly and leave the watering site without delay. They avoid brackish, salty water. They are cryptic and secretive away from water holes, crouching before flushing at close distance when disturbed, or run with a hunched posture through the grass. They have a distinctive, mellow two-note '*chok-lit*' flight call.

Food and feeding

Their diet consists of small, dry seeds of legumes (Fabaceae), with a preference in the Kalahari for *Lophiocarpus burchelli*, *Requernia sphaerosperma* and *Tephrosia* species. They prefer to forage in grassy areas with well-drained, sandy substrata.

Breeding

They are monogamous solitary nesters and breed during the dry season from April to October. The nest is a scrape in the sand, usually placed next to a grass tuft or shrub. The clutch of 3 cream to tawny-olive eggs, with elongated smears of dark olive-brown and greyish-mauve measure 36.7 x 25.7 mm. The female incubates by day and the male at night. The incubation period is unknown.

Status and conservation

They have benefited from the provision of artificial watering sites, such as borehole livestock troughs, which have allowed them to forage further afield. The loss of breeding adults due to wing-shooting during their breeding season, added to big-game hunting and, coincidentally, during Namaqua Sandgrouse *P. namaqua* hunts can increase the mortality of eggs and flightless chicks. Fortunately, they favour habitats different from Namaqua Sandgrouse, and can therefore be excluded from sites where Namaqua Sandgrouse are regularly hunted by providing watering sites within their favoured habitats. Where they do drink at the same site, they also arrive for drinking later than Namaqua Sandgrouse, which can be recognised by their different call and shooting should then cease.

Madagascan Sandgrouse
Ganga masqué

Pterocles personatus

Pterocles personatus Gould, 1843, Majambo Bay, Madagascar

Male Madagascan Sandgrouse (both photos by Dubi Shapiro)

A pair of Madagascan Sandgrouse in flight, female on the right (photo by Otto Schmidt)

Classification

They are monotypic with no subspecies recognised.

Description

- They are short-tailed rufous-tan sandgrouse with black flight feathers.
- Males have a bold black face mask surrounding the bill, a yellow area of bare skin around the eye, a pinkish-buff breast, a light brown mottled back, brown wings and paler underparts barred with black.
- Females are generally duller with tan upperparts, finely speckled and barred with brown.
- Juveniles resemble females but have duller plumage.
- Downy chicks are warm buff above with a buffy-white median band and two transverse bands, and buffy white below, with buffy-white lines above and below the eye.

Distribution and habitat

- They are the only sandgrouse found on Madagascar.
- They are a common endemic to western and south-western Madagascar, commonly seen at the mouth of the Onilahy River near Augustin, at Lake Ranobe and Lake Amboromalandy.
- They inhabit open areas such as grasslands, savanna, open wooded plains and the surrounds of lakes and rivers at low altitudes down to sea level.
- They prefer level, sparsely vegetated ground, sometimes with rocks, including *Xerophyta* savannas, slightly degraded and open spiny forest.

Habits

They are usually found in small groups that fly fast and powerfully when flushed. They drink mainly in the early morning, but may appear at waterholes, including rivers, ponds and lakes, at any time of the day, typically in groups of 2–6, more exceptionally 20–40 individuals, with totals of 300–400 birds reported at favoured sites, particularly in the dry season. The flight-call is a distinctive, low pitched, nasal, repetitive '*katakataka*', '*ak-ak-ak-ak-ak-ak-ak-ak*', which may also be given on the ground.

Food and feeding

They forage in slightly degraded or open spiny forest or areas dominated by *Xerophyta* or introduced grasses. Their diet consists of seeds, and insects are also reported in their diet.

Breeding

They are monogamous solitary nesters, but sometimes nest in loose colonies. They breed through the late dry and early wet season (August–November) and again in the early dry season (April–June). The nest is a scrape in gravelly, sandy or hard ground, often lined with some plant fibres, usually on level or slightly sloping ground, occasionally at the foot of a shrub within relatively dense vegetation or under large *Tamarindus indica* plants, sometimes on lakeshores and even on coastal beaches. The clutch of 3 buff to greenish-brown eggs are finely marked with brown and grey, similar to those of Black-bellied Sandgrouse *P. orientalis*, and measure 45.0 × 32.6 mm. Incubation is by both sexes, but the period is unknown.

Status and conservation

They are not considered to currently have any specific conservation concern largely because they have a wide range within Madagascar and are common in the west and south of the country, although they are less common further north.

PART 6
Snipes & Eurasian Woodcock

SNIPES, GREATER PAINTED-SNIPE AND EURASIAN WOODCOCK

Classification

Snipes are wading birds of the Order Charadriiformes consisting of 25 species globally within three genera:

- *Coenocorypha* Gray, GR, 1855
- *Gallinago* Brisson, 1760
- *Lymnocrytes* Boie, F, 1826, in the family Scolopacidae

The three species of painted-snipes, including the Greater Painted-snipe *Rostratula benghalensis* of Africa, are not closely related to typical snipes, and are placed in their own family, the Rostratulidae.

The Eurasian Woodcock *Scolopax rusticola* is also a member of the family Scolopacidae.

Description

- Snipes are cryptic waders with relatively short legs, a characteristically disproportionate long and slender bill, and superbly camouflaged plumage.
- Unlike true snipes where the males and females appear very similar, painted-snipes have distinct sexual dimorphism, and even more unusual, the females are more striking than the males.

Natural history

Gallinago snipe have almost a worldwide distribution and include three of the African snipes, the genus *Lymnocryptes* includes Jack Snipe *L. minimus*, which occur in Asia and Europe, and are Palearctic migrants to central and West Africa, while *Coenocorypha* snipes are found only in the outlying islands of New Zealand. Pin-tailed Snipe *G. stenura*, which is a migrant between northern Russia and southern Asia, is only a rare vagrant to Socotra, Somalia and East Africa, and is not covered in this book.

Snipes favour various types of wet, marshy habitats including swamps, flooded grasslands, and along muddy rivers and estuaries with emergent vegetation. Snipes generally avoid areas with dense vegetation, preferring marshy areas with patchy mud openings to forage and avoid predators.

PREVIOUS PAGE A pair of Greater Painted-snipe with the female crouched on the right (photo by Andre Botha)

The Eurasian Woodcock is the only one of five woodcock species that visits northern Africa from the western Palearctic as a non-breeding migrant.

Evolutionary placement

- Snipes, painted-snipes and woodcocks are all waders, also known as shorebirds.
- The shorebirds are commonly split into two suborders or clades:
  - Scolopaci, which include snipes, painted-snipes, woodcocks, jacanas, sandpipers, curlews, godwits and dowitchers; and
  - Charadrii, which include plovers (lapwings), stilts, avocets, thick-knees and sheathbills.

Conservation

Their superb camouflage and tendency to stand motionless when approached make them elusive for the bird watcher and wing-shooter to detect, before their explosive flush. When flushed, snipes fly fast, some with a dodging flight path, which leaves the birder baffled and hunters with difficulty estimating a correct aiming lead for the bird's erratic flight pattern. It is suggested that the difficulties involved in hunting snipe gave rise to the term 'sniper'.

The two main threats to their population stability is netting in Africa, Europe and the Middle East, and the degradation and drainage of wetlands. Only two of the seven species covered in this section are threatened; both the Great Snipe *Gallinago media* and Madagascan Snipe *G. macrodactyla* are listed as Near Threatened (The IUCN Red List of Threatened Species v 2015-4).

Greater Painted-snipe
Rhynchée peinte

Rostratula benghalensis

Rallus benghalensis Linnaeus, 1758, Bengal, Asia

Male Greater Painted-snipe (photo by Ian White)

Male Greater Painted-snipe (photo by Ian White)

Female Greater Painted-snipe (photo by Andre Botha)

Two male Greater Painted-snipe (photo by Andre Botha)

Male Greater Painted-snipe at the nest (photo by Andre Botha)

Female Greater Painted-snipe showing her attractive wing pattern (photo by Andre Botha)

Female Greater Painted-snipe landing (photo by Andre Botha)

Classification

They are monotypic with no subspecies recognised. A suggested separate race, *R. b. madagascariensis*, on Madagascar, based on being darker and greyer than mainland African birds is not yet recognised.

Description

- They differ from true snipes by having a white eye-ring which extends onto the ear coverts, a dark breast band and an obvious white breast which extends up onto the shoulders like white braces.
- They have rounded wings and a slow wingbeat.
- Their bill is pale, shorter than that of true snipes and is slightly decurved at the tip.
- The legs are longer than those of true snipes.
- Females are the dominant sex and are more strikingly marked, with a rich chestnut neck and breast, and uniform upperparts.
- Males and juveniles are more cryptic, with a grey neck, buff spotting on the upperparts and a conspicuous golden 'V' on the back.
- Juveniles are similar to the adult male in appearance, but are generally slightly greyer and the breast appears streaked rather than barred.
- Downy chicks are pale with a dark stripe on the crown and down the centre of the back, and another dark stripe across the side of the face and across the upper-wings on each side.

Distribution and habitat

- They have an extensive Old World range, from sub-Saharan Africa and Madagascar east to southern Asia and Japan.
- In Africa, they are patchily distributed from Senegambia across to Egypt and south to South Africa and Madagascar.
- They are absent from forested and desert areas.
- They are generally an uncommon resident and local nomad.
- Favoured habitats are marshes and flooded grasslands, particularly with good vegetation cover and receding water levels, and exposed mud flats between the vegetation.

Habits

They are usually solitary, in pairs, or occasionally in small groups. They are secretive, reluctant to flush, and can remain motionless for long periods. Flight is slow on large, broad wings, with legs often trailing, somewhat rallid-like. Silent when flushed. Females give a wheezy hiss and repeated '*wuk-oooooo*', repeated monotonously in display, often at night.

Food and feeding

Much like true snipes, they forage by probing mud with their bill, frequently bobbing their body. Their diet consists of worms, grasshoppers, crickets, crustaceans, snails and seeds.

Breeding

They are polyandrous, with a female mating with 2–4 males during each breeding season. The nest is a slight depression usually hidden in short sedges or grasses, with vegetation bent and trodden down to form the nest. The clutch of 2–5 eggs measure 35.5 x 25.0 mm, and are smooth creamy-beige with bold scrolls and blotches of dark brown and black. Incubation by the male only lasts for 17–19 days. Chicks are cared for by the male only and fledge at 30–35 days.

Status and conservation

Although there is no evidence of significant changes in distribution, local populations can be severely affected by rural development. Their favoured habitats can also be transformed by dam construction, wetland modification and the invasion of flood plains by tall vegetation, for example, Bulrushes *Typha capensis*, which interfere with streamflow and seasonal flooding cycles.

Jack Snipe
Bécassine sourde

Lymnocryptes minimus

Scolopax minima (Br nnich, 1764), Christiansø Islands, off Bornholm, Denmark

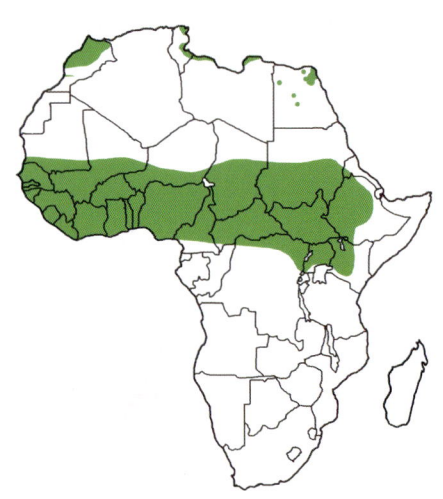

Jack Snipe (photo by Tasso Leventis)

Jack Snipe (photo by Markus Lagerqvist)

Classification

The Jack Snipe is monotypic with no subspecies recognised.

Description

- They are smaller and shorter-billed than the Common Snipe *G. gallinago*, with narrow wings and a buffy hind-neck.
- Unlike other African snipes, they have a distinctive head pattern where the crown lacks a pale central stripe, rather they have a double, pale buff supercilium which splits above the eye.
- They have two bold yellowish stripes on each side of the dark glossy upperparts and streaked breast and flanks.
- The upper tail also differs by being wedge-shaped and wholly dark.
- Juveniles are similar to adults but have white under-tail coverts with smaller and paler brown stripes.

Distribution and habitat

- They are an uncommon Palearctic migrant to the northern Afrotropics and Morocco, visiting the continent mostly from October to March.
- They are also a rare vagrant to Bird Island, Seychelles.
- They prefer brackish and freshwater habitats including moist and waterlogged areas with soft silty mud and short grass or sedges.

Habits

They are mostly crepuscular or nocturnal. They flush less readily than other snipes, preferring to freeze and rely on their camouflage. Their flight is also relatively slow and weak, less erratic, and they drop down into cover sooner. Generally silent in the non-breeding season, but when flushed, sometimes utter a faint '*gah*'.

Food and feeding

They usually feed singly or in small loose groups. When foraging they have a distinctive rhythmic bobbing motion while probing in soft mud or picking up food items from the surface. Their diet consists mostly of adult and larval insects, annelids, small freshwater and terrestrial gastropods and sometimes seeds or other plant material.

Breeding

They are a non-breeding migrant to Africa, breeding mostly in Scandinavia, Russia and Siberia in wet marshy bogs, tundra and meadows during the boreal summer.

Status and conservation

They are not globally threatened although populations have declined in some areas. The major threats are wetland degradation and drainage for agricultural intensification, afforestation and peat extraction. Hunting pressure is also a threat, where during the autumn migration, 5% of the European population is believed to be shot annually.

African Snipe
Bécassine africaine

Gallinago nigripennis

Gallinago nigripennis Bonaparte, 1839, Cape of Good Hope, South Africa

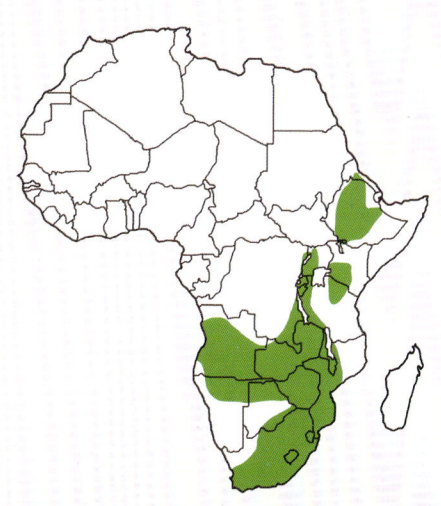

African Snipe with sediment covered bill
(photo by Ian White)

Camouflaged African Snipe
(photo by Jacques Pitteloud)

African Snipe on frosty vegetation
(photo by Andre Botha)

Landing African Snipe
(photo by Andre Botha)

Classification

There are three subspecies recognised:

G. n. nigripennis Bonaparte, 1839 in southern Mozambique and South Africa

G. n. aequatorialis Rüppell, 1845 in Ethiopia to eastern Democratic Republic of the Congo, eastern Zimbabwe and northern Mozambique

G. n. angolensis Bocage, 1868 in Angola and northern Namibia across to western Zimbabwe and Zambia

Description

- The African Snipe, also known as Ethiopian Snipe in some parts, is darker on top than the Common Snipe *G. gallinago*, and has heavily streaked neck and breast feathers, which contrast more with the white belly.
- In flight, they have short broad wings and also show more white in the outer-tail feathers, even though the five outer-tail feathers are barred black and white.
- The pale fringes to the greater coverts contrast more with the darker wings than for that of the Great Snipe *G. media*, the wings are shorter and broader and the bill is slightly longer.
- Juveniles are similar to adults, but are less strongly marked on the body and the wings are more mottled and barred.
- Downy chicks are rich tawny with a black spot above the base of the bill, a black streak through the lores and chin, and a band across the forehead.

Distribution and habitat

- African Snipe are common residents with varying local nomadism.
- They occur from central Ethiopia, down the Rift Valley, across to Angola and down to South Africa, with a separate population in western Kenya and northern Tanzania.
- Their favoured habitats include marshes and flooded grasslands, usually in muddy areas with short, patchy vegetation.

Habits

They usually occur as scattered individuals, but many can inhabit one wetland. If alarmed, they freeze, and their cryptic plumage provides effective camouflage when the bird stands motionless among wetland vegetation. When flushed they give a sucking 'scaap' sound and their flight is more erratic than that of Great Snipe. Males make a whirring '*whu-whu-whu-whu-whu*' drumming sound with their stiffened outer-tail feathers during their stooping aerial display flights.

Food and feeding

They are mainly crepuscular or nocturnal foragers, probing soft mud for prey. Their diet consists of annelids, the larvae of beetles, dragonflies and flies, small molluscs and crustaceans.

Breeding

They are monogamous solitary nesters, breeding mostly after the rainy season. The nest is a small platform of grass within vegetation tussocks surrounded by wet or flooded ground. The clutch of 2–3 pale greenish or buffish eggs are strongly spotted and blotched blackish-brown, pale brown and slate grey mostly at the large end, and measure 41.2 x 29.6 mm.

Status and conservation

They are threatened by habitat loss and degradation, such as decreases in the area of natural pastures and lake-edges and stream-edge marshes, agricultural expansion and increases in the area of lake mudflats due to increased siltation and reduced water supply because of bush encroachment, overgrazing, burning and the draining of wetlands.

Great Snipe
Bécassine double

Gallinago media

Scolopax media (Latham, 1787), England

Great Snipe in a cropland in Lubumbashi, Kataga, Democratic Republic of the Congo (photo by Stéphane Doppagne)

Great Snipe (photo by Tasso Leventis)

Great Snipe flight patterns (photo by Bruno Portier)

Great Snipe (photo by Stéphane Doppagne)

Classification
The Great Snipe is monotypic with no subspecies recognised.

Description
- They are most easily distinguished from the other African snipes by their boldly barred flanks, belly and vent.
- Adults also have large, white tips to trailing upper-wing coverts forming prominent white wing bars and conspicuous white sides and tip of the outer-tail feathers.
- They differ from the Common Snipe *G. gallinago* by being slightly larger with a shorter bill, their underparts are more heavily marked, and they appear darker and heavier in flight, with a slower, more direct flight.

Distribution and habitat
- The Great Snipe is an uncommon Palearctic migrant to the flatter, open regions of West Africa, and from Sudan and Ethiopia in north-central Africa down to the northern parts of southern Africa, eastern South Africa and the Seychelles.
- They are most common in Africa from October to May.
- Similar to other snipes, their favoured habitats in Africa include marshes and flooded grasslands, but can also be found in short grass away from water.

Habits
They usually occur singly or in small loose groups. When flushed, they fly fast in a straight line, not dodging and jinking like the flight of African Snipe *G. nigripennis* and Common Snipe. Their flush flight is also usually fairly short. The bill is held more level in flight. They are generally silent, but occasionally utter one or two short croaks when flushed, deeper than that of the African Snipe.

Food and feeding
They are crepuscular or nocturnal foragers, feeding by probing in soft mud much like the other snipes. Their diet consists mainly of annelids, small molluscs, insects and some plant material.

Breeding
They are a non-breeding migrant to Africa, breeding mostly in Scandinavia and Russia in wet marshy bogs, tundra and meadows during the boreal summer.

Status and conservation

They are experiencing a population decline, owing primarily to habitat loss in their breeding range, including conversion of marshlands and floodplain meadows to intensive agriculture, wetland drainage and the submergence of river valleys during the creation of dams, as well as to hunting in eastern Europe and in their African wintering range. Their conservation status is therefore listed as Near Threatened (The IUCN Red List of Threatened Species v 2015-4).

Common Snipe
Bécassine des marais
Gallinago gallinago
Scolopax gallinago (Linnaeus, 1758), Sweden

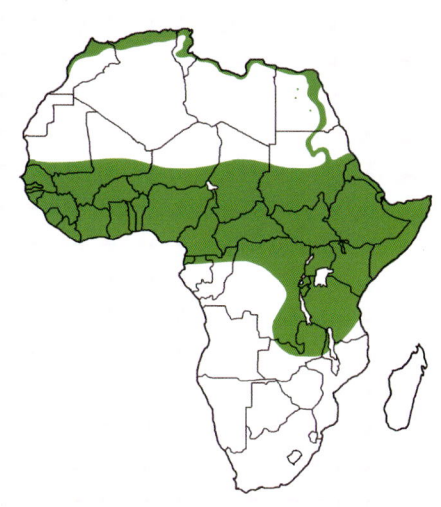

Common Snipe fan-tail display (photo by Markus Lagerqvist)

Common Snipe showing the specialised tail feathers for the drumming display (photo by Tasso Leventis)

A moving flock of Common Snipe (photo by Tasso Leventis)

Common Snipe (photo by Tasso Leventis)

Classification

There are two subspecies recognised:

G. g. faeroeensis (Brehm, CL, 1831) in Iceland, Faroe Islands, Orkney and Shetland Islands

G. g. gallinago (Linnaeus, 1758) in Europe, Asia and Africa

Description

- The Common Snipe is similar to the African Snipe *G. nigripennis*, but is slightly paler above and has much less white in the outer tail, with the white confined to the tail tips and outer margins of the outer tail feathers.
- Their bill is slightly shorter, and the wings are longer and more pointed.
- Another similar species is the Pin-tailed Snipe *G. stenura*, which is a rare vagrant to Socotra, Somalia and East Africa, not covered in this book. However, the Pin-tailed Snipe has a narrow pale grey trailing edge to the secondaries in flight, which is obvious and white in Common Snipe.
- Furthermore, the toes of the Pin-tailed Snipe project slightly beyond the tail tip in flight.
- The Pin-tailed Snipe is also slightly paler and has a broader pale eye-stripe from the base of the bill.

Distribution and habitat

- They are a common Palearctic migrant to Africa, presumably from Russia, during October to March and also a vagrant to the Seychelles.
- They are common across the northern Afrotropics down to the northern Democratic Republic of the Congo and northern Tanzania, less common in northern Zambia and Malawi.
- They prefer permanent and perennial swamps, marshy edges of lakes and dams, and seasonally flooded grasslands and rice fields.

Habits

Although largely solitary, they will form small groups. They are mainly crepuscular, resting in wetland vegetation during the day. They walk deliberately with the bill angled downwards. They squat when disturbed, but, if flushed, they have an explosive, zig-zag tilting flight, and utter a loud, harsh '*creech*', which is more rasping than that of the African Snipe. They will also perch on a low object while uttering an insistent '*chip-per, chip-per*'.

Food and feeding

They forage in shallow water sometimes up to their belly in the water, probing deeply and vigorously in soft mud, often immersing their bill up to eye level. Their diet consists of annelids, insects and their larvae, small molluscs and crustaceans, spiders, small frogs and some plant material.

Breeding

They are a non-breeding migrant to Africa, breeding mostly in the Azores, Iceland and widely through Europe, Russia and North America in wet marshy bogs, tundra and meadows during the boreal summer.

Status and conservation

They are generally common to abundant across the northern Afrotropics and are not globally threatened. The drainage or degradation of wetlands limits their foraging habitat. They are hunted fairly heavily in Europe.

Madagascan Snipe
Bécassine malgache
Gallinago macrodactyla

Gallinago macrodactyla Bonaparte, 1839, Madagascar

Madagascan Snipe from Torotorofotsy Marsh, Madagascar (photo by Louise Jasper)

Madagascan Snipe from Vohiparara, Ranomafana, Madagascar (photo by Margaret Sloan)

Classification

Madagascan Snipe are monotypic with no subspecies recognised. They have previously been regarded as conspecific with Common Snipe *G. gallinago* and Noble Snipe *G. nobilis*, but they are resident in Madagascar.

Description

- Their relatively long straight bill and striped head are diagnostic.
- They are similar to African Snipe *G. nigripennis*, but with less contrast between the upperparts and the underparts, and their outer rectrices are greyer brown.
- They are also similar to Common Snipe and other close relatives, but they have a longer dark bill and their legs are dark yellow-olive. They have a narrow greyish trailing edge to secondaries, darker underwings and wing-coverts with warm olive-buff fringes.
- Females are very similar to males, but their average bill length is slightly longer.
- Juveniles are similar to adults but have narrower and paler warm olive-buff fringes to scapulars and mantle feathers.

Distribution and habitat

- They are the only snipe found on Madagascar.
- They are endemic to the eastern and central massif of Madagascar where they are common in the east and north-west.
- They are regularly seen at the Torotorofotsy marshes near Perinet-Analamazaotra and at the marshes and wetlands around Vohiparara in Ranomafana.
- They inhabit largely freshwater, grassy and sedge-covered marshes and swamps, flooded fields, muddy margins of water courses with dense vegetation and rice paddies, from sea level up to elevations of 2 700 m above sea level.

Habits

They are sedentary and secretive. Their flight is fast with a zig-zagging flight path. Their drumming in the courtship display is a noisy '*woo-woo-woo*' during the breeding season. When flushed, they give a harsh rapidly repeated '*eck-eck-eck-eck*' call.

Food and feeding

Outside the breeding season, they frequently forage in small groups of 4-8 birds, probing deep into mud on muddy flats or in shallow water. Their diet consists of insects and worms, seeds and other plant matter.

Breeding

They breed in wet marshy habitats mostly from July to January. The nest is a saucer-shaped depression scantily lined with dry grass, usually on a dry vegetation hammock covered with grass or sedges. The clutch consists of 2–4 eggs, but usually 3. No further information is available on their breeding biology.

Status and conservation

They are threatened by habitat transformation and drainage of natural wetlands, particularly with the spread of rice paddy fields, and hunting pressure may also be a threat. Their conservation status is therefore listed as Near Threatened (The IUCN Red List of Threatened Species v 2015-4).

Eurasian Woodcock
Bécasse des bois

Scolopax rusticola

Scolopax rusticola Linnaeus, 1758, Sweden

Flushing Eurasian Woodcock
(photo by Eduardo Gutierrez González)

Eurasian Woodcock
(photo by Andrew Hoodless)

Two Eurasian Woodcocks
(photo by Andrew Hoodless)

Eurasian Woodcock in flight
(photo by Eduardo Gutierrez González)

GAMEBIRDS

Classification

The Eurasian Woodcock is a member of the family Scolopacidae, along with the snipes. They are monotypic with no subspecies recognised.

Description

- They are a bulky snipe-like bird with an average weight of 300 g.
- They have cryptic camouflage to suit their woodland habitats, with intricately patterned reddish-brown upperparts and buff underparts.
- Their head is barred with black, not striped like that of its close relatives – the snipes.
- They have large eyes located high on the sides of their head.
- The robust straight bill is flesh brown and darker at the tip.
- The sexes are similar and their heavy build, long broad rounded wings and slow wing-beat differentiate them from the snipes.

Distribution and habitat

- Their distribution is widespread across temperate and subarctic Eurasia, from western Europe to Russia and Siberia.
- They are a Palearctic wintering visitor to north-western Africa, arriving in October and departing in March.
- They inhabit damp open woodlands, coastal areas with thick cover and well vegetated gardens.

Habits

They are crepuscular and rarely active during the day. When flushed, they fly off with a whirring wing noise. They fly direct while migrating or crossing open country, but fly erratically with twisting and fluttering once in woodland. They are usually solitary and migrate singly, but may congregate where foraging conditions are optimal or to roost. When flushed they give a harsh snipe-like '*schaap*'.

Food and feeding

They forage by walking and probing deeply in soft marshy soil in thickets, usually well hidden from sight. They mainly eat earthworms, but also insects and their larvae, slugs, spiders, freshwater molluscs, some plant seeds and grass roots.

Breeding

They are a non-breeding migrant to Africa. In the breeding grounds, they nest on the ground in low cover in woodland or tall heather.

Status and conservation

In many countries, Eurasian Woodcock are hunted as game, and their size, speed and flight pattern make them a very challenging quarry. Recent research in the United Kingdom and France have disclosed results which call for caution in the harvesting of woodcock populations, particularly wintering in western France and Italy, and could be a forewarning of a decline. Other threats are degradation of habitats and harmful pesticides.

GLOSSARY

Antiphonal	Sung in alternation; calling involving two duetting birds, each giving components of what sounds like a single call.
Aviculturists	People who captive-breed birds for financial gain or as a hobby.
Bib	Rounded patch of feathers covering the lower throat and upper breast of a bird, of a colour contrasting with that of the throat and breast.
Belly-wetting	During the breeding season, male sandgrouse soak up water in specially designed belly feathers to transport the water to flightless chicks in the desert.
Casque	Bony, horn-like growth covered with a sheath of keratin on the head (or upper mandible) of some birds.
Corm	A nutritious underground storage organ of certain geophyte plants.
Covey	A group of francolins or spurfowls, normally one or more family groups comprising a territorial adult male and female and their grown offspring from the previous breeding season.
Crepuscular	Mostly active at dawn and dusk.
Dominance hierarchy	'Pecking order' dominance among individuals, with the dominant bird being the victor in aggressive encounters or the one getting preferred access to resources, such as mates, food and roosts.
Dust-bathe	Sitting and scratching in light, sandy soil which is worked into the feathers and expelled, presumably to remove parasites and debris in the feathers.
Egg-dumping	When one hen lays her eggs in another's nest.
Eye-stripe	Contrasting (dark or light) stripe on the side of the head running backwards above (supercilium) or at the same level as the eye.
Flight call	Call given as birds flush from cover.

Flush	Explosive flight from cover, normally employed only as a last resort when a bird is trying to evade danger.
Fynbos	Mountain fynbos, lowland fynbos: vegetation types, including proteas, ericas and restios, characteristic of the winter rainfall region, mediterranean-like elements of the Cape Floral Kingdom.
Galliform	The order Galliformes includes heavy-bodied, ground-feeding, fowl-like or chicken-like birds.
Gregarious	Regularly associating in groups.
Gular fluttering	Rapid oscillation of the thin floor of the mouth and upper throat, done by heat-stressed birds for evaporative cooling to lower body temperature.
Half-collar	Dark band down each side of the neck and across the breast, but not completely joining on the breast.
Hypothermia	Loss of control of body temperature under very cold and/or wet conditions.
Igneous (igneous substrata)	Soil types derived from volcanic rock.
Incubation	Process whereby adult birds apply body heat to eggs, providing the appropriate temperature and humidity necessary for development of the embryo within.
Irruptive	Rapid, massive influx, or reproduction, of a given bird species leading to large, sudden increases in its population.
Keets	Guineafowl chicks.
Microclimate	Fine-scale differences in temperature and/or humidity, e.g. within the nest under an incubating adult; repeated, cyclic climatic changes over hundreds and/or thousands of years that allow forest and savanna/grassland vegetation to expand and contract.
Migratory	Regular, predictable return movements over the landscape, usually on an annual basis, from a repeatedly used breeding ground to an area some distance away which may provide seasonally abundant food or some other key resource.

Miombo woodland	Mix of savanna and woodland dominated by trees such as *brachystegia*, *combretum*, *terminalia* and *burkea* species and grasses such as *aristida*, *digitaria* and *eragrostis* species.
Monogamous	Mating system in which one adult male mates with only one adult female. See polygynous.
Moustache patch	Strip of distinctively coloured feathers running from the base of the upper bill down the side of the head to below the eye.
Multiple-broods hypothesis	Suggestion that a breeding pair can lay and successfully rear chicks from more than one clutch of eggs in one breeding season.
Nomadic	Descriptive of irregular, unpredictable movements over the landscape, probably to exploit ephemerally abundant food resources.
Offtake levels	Percentage of a population of gamebirds that is removed by wing-shooting or other forms of hunting; ideally such levels should not exceed those that can be sustained by natural reproduction year after year.
Pair-bond	The attraction, or commitment, that keeps a pair of breeding adults together.
Polygynous	Mating system where one adult male mates with more than one adult female.
Precocial	Descriptive of newly hatched young that are active and leave the nest immediately after hatching, and are therefore not fully dependent on their parents.
Resident	Descriptive of a bird that spends its entire life at or near the area in which it hatched or where it has established a territory, without seasonal movement.
Savanna	Vegetation with a mixture of grass and some bush and trees, often maintained by fire or feeding by elephants.
Sedentary	Tending to remain near a particular area (e.g. the site of hatching) for life, with minimal dispersal.
Sexual dimorphism	In which males and females of the same species differ markedly in their plumage and/or size.

Species	The fundamental products of evolution and units of conservation management; often identified by evidence of a lack of interbreeding with very similar birds occurring on the same ground.
Subspecies	Geographical races of a single species between which there are no intrinsic barriers to interbreeding.
Supercilium	A plumage stripe which runs from the base of a bird's beak above its eye, ending towards the rear of the bird's head.
Territorial	Process of defending a piece of ground from the intrusion of members of the same species; normally by males only.
Vermiculated	Refers to irregular, ill-defined, 'wormy' barring.
Wing-shooting	Hunting of gamebirds in flight with shotguns; pointing dogs locate the birds and retrievers collect shot birds.

BIBLIOGRAPHY AND FURTHER READING

(References listed in *Gamebirds of Southern Africa* (Little and Crowe 2011) have not been listed here)

Awa. T. II, Tamungang, S.A., Marcel, A.E. and Champlain, D.L. 2015. Preliminaries on population status and distribution of the endangered Mount Cameroon Francolin, *Pternistis camerunensis*. *Newsletter of the IUCN-SSC Galliformes Specialist Group* 10: 13–17.

Borrow, N. and Demey, R. 2004. *Birds of Western Africa*. London: Christopher Helm.

Borrow, N. and Demey, R. 2011. *Birds of Senegal and The Gambia*. London: Christopher Helm.

Borrow, N., Demey, R., Owusu, E.H. and Ntiamoa-Baidu, Y. 2010. *Birds of Ghana*. London: Christopher Helm.

Bowie, R.C.K. and Fjeldså, J. 2005. Genetic and morphological evidence for two species in the Udzungwa forest partridge. *Journal of East African Natural History* 94: 191–201.

Crowe, T. 2016. African gamebirds are the key to understanding global avian evolution. *The Conversation* 24 February. (retrieved from www.theconversation.com/african-gamebirds-are-the-key-to-understanding-global-avian-evolution-54970 on 11 March 2016).

Del Hoyo, J., Elliott, A. and Sargatal, J. (eds). 1994. *Handbook of the Birds of the World*. Vol. 2. Barcelona: Lynx Edicions.

Del Hoyo, J., Elliott, A., Sargatal, J., Christie, D.A. and de Juana, E. (eds). 2013. *Handbook of the Birds of the World Alive*. Barcelona: Lynx Edicions (retrieved from www.hbw.com/node/54091 between July 2015 and March 2016).

Dinesen, L., Lehmberg, T., Vendsen, J., Hansen, L.A. and Fjeldså, J. 1994. A new genus and species of perdicine bird (Phasianidae, Perdicini) from Tanzania; a relict form with Indo-Malayan affinities. *Ibis* 136: 2–11.

Dowsett, R.J., Aspinwall, D.R. and Dowsett-Lemaire, F. 2008. *The Birds of Zambia. An Atlas and Handbook*. Tauraco Press and Aves a.s.b.l. Liège, Belgium.

Gill, F. and Donsker, D. (eds). 2015. *IOC World Bird List* (v 5.3). doi: 10.14344/IOC.ML.5.3. (retrieved from www.worldbirdnames.org between July 2015 and March 2016).

Hall, B.P. 1963. The francolins, a study in speciation. *Bulletin of the British Museum of Natural History (Zoology)* 10: 105–204.

Hockey, P.A.R., Dean, W.R.J. and Ryan, P.G. (eds). 2005. *Roberts – Birds of Southern Africa*, VIIth ed. Cape Town: The Trustees of the John Voelcker Bird Book Fund.

Little, R.M. and Crowe, T.M. 2011. *Gamebirds of Southern Africa*. Cape Town: Struik Nature.

Mandiwana-Neudani, T.G. 2013. Taxonomy, phylogeny and biogeography of francolins ('*Francolinus*' spp.) Aves: Order Galliformes Family: Phasianidae. PhD thesis, University of Cape Town (retrieved from www.open.uct.ac.za/handle/11427/9295 on 30 June 2015).

Mandiwana-Neudani, T.G., Bowie, R.C.K., Hausberger, M., Henry, L., Crowe, T.M. and Craig, A.J.F.K. 2014. Taxonomic and phylogenetic utility of variation in advertising calls of francolins and spurfowls (Galliformes: Phasianidae), *African Zoology*, 49(1): 54–82.

Mandiwana-Neudani, T.G., Kopuchian, C., Louw, G. and Crowe, T.M. 2011. A study of gross morphological and histological syringeal features of true francolins (Galliformes: *Francolimls, Scleroptila, Peliperdix* and *Dendroperdix* spp.) and spurfowls *(Ptemistis* spp.) in a phylogenetic context, *Ostrich* 82: 115–127.

Morris, P. and Hawkins, F. 1998. *Birds of Madagascar*. East Sussex: Pica Press.

Redman, N., Stevenson, T., Fanshawe, J., Borrow, N. and Small, B.E. 2009. *Birds of the Horn of Africa: Ethiopia, Eritrea, Djibouti, Somalia and Socotra*. London: Christopher Helm.

Sinclair, I. and Langrand, O. 2013. *Birds of the Indian Ocean Islands*. Cape Town: Struik Nature.

Sinclair, I. and Ryan, P. 2010. *Birds of Africa South of the Sahara*. Cape Town: Random House Struik.

Stevenson, T. and Fanshawe, J. 2002. *Field Guide to the Birds of East Africa*. London: T & AD Poyser.

The IUCN Red List of Threatened Species. Version 2015-4. (retrieved from www.iucnredlist.org on 17 May 2016).

Töpfer, T., Podsiadlowski, L. and Gedeon, K. 2014. Rediscovery of the Black-fronted Francolin *Pternistis (castaneicollis) atrifrons* (Conover, 1930) (Aves: Galliformes: Phasianidae) with notes on biology, taxonomy and conservation. *Vertebrate Zoology* 64: 261–271.

Urban, E.K., Fry, C.H. and Keith, S. (eds). 1986. *The Birds of Africa*, Vol. II. London: Academic Press.

INDEX

ENGLISH COMMON NAMES

Francolin
- Archer's 79
- Coqui 60
- Crested 57
- Finsch's 85
- Forest (Latham's) 94
- Grey-winged 72
- Moorland 88
- Orange River 76
- Red-winged 82
- Ring-necked 91
- Schlegel's 66
- Shelley's 69
- White-throated 63

Guineafowl
- Black 32
- Crested 43
- Helmeted 35
- Plumed 40
- Vulturine 46
- White-breasted 29

Painted-snipe
- Greater 270

Partridge
- Barbary 97
- Madagascan 109
- Nahan's 116
- Rubeho Forest 106
- Sand 100
- Stone 112
- Udzungwa Forest 103

Peafowl
- Congo 49

Quail
- Blue 215
- Common 208
- Harlequin 212

Sandgrouse
- Black-bellied 236
- Black-faced 245
- Burchell's 259
- Chestnut-bellied 229
- Crowned 242
- Double-banded 255
- Four-banded 252
- Lichtenstein's 249
- Madagascan 263
- Namaqua 226
- Pin-tailed 223
- Spotted 233
- Yellow-throated 239

Snipe
- African 277
- Common 283
- Great 280
- Jack 274
- Madagascan 286

Spurfowl
- Ahanta 164
- Black-fronted 149
- Cape 200

Chestnut-naped 146
Clapperton's 179
Djibouti 143
Double-spurred 173
Erckel's 140
Grey-breasted 137
Grey-striped 170
Handsome 155
Hartlaub's 190
Harwood's 193
Hildebrandt's 182
Hueglin's 176
Jackson's 152
Mount Cameroon 158
Natal 185
Red-billed 196
Red-necked 125
Scaly 167
Swainson's 129
Swierstra's 161
Yellow-necked 133

Woodcock
Eurasian 289

FRENCH COMMON NAMES

Bécasse
des bois 289

Bécassine
Africaine 277
des marais 283
double 280
malgache 286
soured 274

Caille
Arlequine 212
Bleue 215
de Madagascar 109
des blés 208

Francolin
à ailes grises 72
à bandes grises 170
à bec jaune 176
à bec rouge 196
à collier 91
à cou jaune 133
à cou roux 146
à double éperon 173
à front noir 149
à gorge blanche 63
à gorge rouge 125
à poitrine grise 137
coqui 60
criard 200
d'Ahanta 164
d'Archer 79
d'Erckel 140
de Clapperton 179
de Finsch 85
de Hartlaub 190
de Harwood 193
de Hilderbrandt 182
de Jackson 152
de Latham 94
de Levaillant 82
de L'Orange 76
de Nahan 116
de Shelley 69
de Schlegel 66
de Swainson 129
de Swierstra 161
des Somalis 143
du Mont Cameroun 158
du Natal 185
écaillé 167
huppé 57
montagnard 88
noble 155

Ganga
    à face noire 245
    à gorge jaune 239
    à ventre brun 229
    bibande 255
    cata 223
    couronné 242
    de Burchell 259
    de Lichtenstein 249
    masque 263
    Namaqua 226
    Quadribande 252
    tacheté 233
    unibande 236

Paon
    du Congo 49

Perdrix
    de Hey 100
    gambra 97

Pintade
    à poitrine blanche 29
    de Purcheran 43
    noire 32
    plumifère 40
    sauvage 35
    vulturine 46

Poulette
    de rocher 112

Rhynchée
    Peinte 270

Xénoperdrix
    des Rubeho 106
    de Tanzanie 103

SCIENTIFIC NAMES

Acryllium
    vulturinum 46

Afrocolinus
    lathami 94

Afropavo
    congensis 49

Agelastes
    meleagrides 29
    niger 32

Alectoris
    barbara 97

Ammoperdix
    heyi 100

Coturnix
    coturnix 208
    delegorguei 212

Dendroperdix
    sephaena 57

Gallinago
    gallinago 283
    macrodactyla 286
    media 280
    nigripennis 277

Guttera
    plumifera 40
    pucherani 43

Lymnocryptes
    minimus 274

Margaroperdix
    madagarensis 109

Numida
    meleagris 35

Peliperdix
    albogularis 63
    coqui 60
    schlegelii 66

Pternistis
    adspersus 196
    afer 125
    ahantensis 164
    atrifrons 149

bicalcaratus 173
camerunensis 158
capensis 200
castaneilcollis 146
clappertoni 179
erckelii 140
griseostriatus 170
hartlaubi 190
harwoodi 193
hilderbrandti 182
icterorhynchus 176
jacksoni 152
leucoscepus 133
natalensis 185
nobilis 155
ochropectus 143
rufopictus 137
squamatus 167
swainsonii 129
sweirstrai 161

Pterocles
alchata 223
bicinctus 255
burchelli 259
coronatus 242
decorates 245
exustus 229
gutturalis 239
lichtensteinii 249
namaqua 226
orientalis 236
personatus 263
quadricinctus 252
senegallus 233

Ptilopachus
nahani 116
petrosus 112

Rostratula
benghalensis 270

Scleroptila
afra 72
finschi 85
gutturalis 79
levaillantii 82
levalliantoides 76
psilolaema 88
shelleyi 69
streptophora 91

Scolopax
rusticol 289

Synoicus (Excalfactoria)
adansonii 215

Xenoperdix
obscuratus 106
udzungwensis 103

First published by Jacana Media (Pty) Ltd in 2016

10 Orange Street
Sunnyside
Auckland Park 2092
South Africa
+2711 628 3200
www.jacana.co.za

© Rob Little, 2016
© Distribution maps Bryan Little, 2016

All rights reserved.

ISBN 978-1-4314-2414-6

Cover design by Shawn Paikin
Set in Sabon 9.5/13pt
Printed by Creda Communications
Job no. 002751

See a complete list of Jacana titles at www.jacana.co.za

FRONT COVER Mount Cameroon Spurfowl (photo by Hadoram Shirihai)

BACK COVER Pin-tailed Sandgrouse in flight (photo by Francois Mougeot)